Integrated Disaster Risk Management

Just the first one and one-half decades of this new century have witnessed a series of large-scale, unprecedented disasters in different regions of the globe, both natural and human-triggered, some quite new and others conventional. Unfortunately, this adds to the evidence of the urgent need to address such crises as time passes. It is now commonly accepted that disaster risk reduction (DRR) requires tackling the various factors that influence a society's vulnerability to disasters in an integrated and comprehensive way, and with due attention to the limited resources at our disposal. Thus, integrated disaster risk management (IDRiM) is essential. Success will require integration of disciplines, stakeholders, different levels of government, and of global, regional, national, local, and individual efforts. In any particular disaster-prone area, integration is also crucial in the long-enduring processes of managing risks and critical events before, during, and after disasters.

Although the need for integrated disaster risk management is widely recognized, there are still considerable gaps between theory and practice. Civil protection authorities; government agencies in charge of delineating economic, social, urban, or environmental policies; city planning, water and waste-disposal departments; health departments, and others often work independently and without consideration of the hazards in their own and adjacent territories or the risk to which they may be unintentionally subjecting their citizens. Typically, disaster and development tend to be in mutual conflict but should, and could, be creatively governed to harmonize both, thanks to technological innovation as well as the design of new institutions.

Thus, many questions on how to implement integrated disaster risk management in different contexts, across different hazards, and interrelated issues remain. Furthermore, the need to document and learn from successfully applied risk reduction initiatives, including the methodologies or processes used, the resources, the context, and other aspects are imperative to avoid duplication and the repetition of mistakes.

With a view to addressing the above concerns and issues, the International Society of Integrated Disaster Risk Management (IDRiM) was established in October 2009.

The main aim of the IDRiM Book Series is to promote knowledge transfer and dissemination of information on all aspects of IDRiM. This series will provide comprehensive coverage of topics and themes including dissemination of successful models for implementation of IDRiM and comparative case studies, innovative countermeasures for disaster risk reduction, and interdisciplinary research and education in real-world contexts in various geographic, climatic, political, cultural, and social systems.

More information about this series at http://www.springer.com/series/13465

Katsuya Yamori
Editor

Disaster Risk Communication

A Challenge from a Social Psychological Perspective

Springer

Editor
Katsuya Yamori
Disaster Prevention Research Institute
Kyoto University
Uji, Kyoto, Japan

ISSN 2509-7091 ISSN 2509-7105 (electronic)
Integrated Disaster Risk Management
ISBN 978-981-13-2317-1 ISBN 978-981-13-2318-8 (eBook)
https://doi.org/10.1007/978-981-13-2318-8

This Springer imprint is published by the registered company Springer Nature Singapore Pte Ltd.
The registered company address is: 152 Beach Road, #21-01/04 Gateway East, Singapore 189721, Singapore

Foreword to the IDRiM Book Series

In 2001, the International Institute for Applied Systems Analysis (IIASA) and the Disaster Prevention Research Institute (DPRI) joined hands in fostering a new, interdisciplinary area of integrated disaster risk management. That year, IIASA and DPRI initiated the IIASA–DPRI Integrated Disaster Risk Management Forum Series, which continued over 8 years, helping to build a scholarly network that eventually evolved into the formation of the International Society for Integrated Disaster Risk Management (IDRiM Society) in 2009. The launching of the society was promoted by many national and international organizations.

The volumes in the IDRiM Book Series are the continuation of a proud tradition of interdisciplinary research on integrated risk management that emanates from many scholars and practitioners around the world. In this foreword, we briefly summarize the contributions of some of the pioneers in this field. We have endeavored to be inclusive but realize that we have probably not identified all those worthy of mention. This foreword is not meant to be comprehensive but rather indicative of major contributions to the foundations of IDRiM. This research area is still in a continuous process of exploration and advancement, several of the outcomes of which will be published in this series.

Japan

Disaster Prevention Research Institute

The idea of framing disaster prevention in risk management terms was still embryonic even among academics in Japan when Kobe and its neighboring region were shaken by the Great Hanshin–Awaji Earthquake (GHQ) in 1995. For example, Okada (1985) established the importance of introducing a risk management approach to reduce flood and landslide disaster risks. Additionally, it was not until late 1994 that the Disaster Prevention Research Institute (DPRI) of Kyoto University

Table 1 Conventional disaster plan vs. 21st century integrated disaster planning and management

Reactive	Proactive
Emergency and crisis management	Risk mitigation plus preparedness approach
Countermeasure manual approach	Anticipatory/precautionary approach
Pre-determined planning (if known events)	Comprehensive policy-bundle approach
Sectoral countermeasure approach	Adaptive management approach
Top-down approach	Bottom-up approach

had reorganized to add a new cross-disciplinary division of Sogo Bosai, or "integrated disaster management."

The new division of DPRI undertook a strong initiative among both academics and disaster prevention professionals to substantiate what is meant by integrated disaster management and to communicate to society why it is needed and how it helps. Many of these efforts were based on evidence and lessons learned from the GHQ. Japan's disaster planning and management policy changed significantly thereafter. Table 1 contrasts the approaches before and after that cataclysmic event. The current approach stresses strategies that are proactive, anticipatory, precautionary, adaptive, participatory, and bottom-up. The rationale is that governments in Japan had been found to be of relatively little help immediately after a high-impact disaster. Lives in peril had more often been saved by the actions of individuals and community residents than by official governmental first responders.

To understand a significant change in disaster planning and management in Japan, one must understand the contrasts among Kyojo ("neighborhood or community self-reliance"), Jijo ("individual or household self-reliance"), and Kojo ("government assistance"). Realizing limitations in the government's capacity after a large-scale disaster, Japan has shifted more toward increasing both Kyojo and Jijo self-reliance roles, and to depend less on the former, which in the past was the major agent to mitigate disasters.

One of the additional lessons learned after the 1995 disaster was to address the need for a citizen-led participatory approach to disaster risk reduction before disasters, as well as for disaster recovery and revitalization after disasters.

International Collaboration

In 2001, the International Institute for Applied Systems Analysis (IIASA) and DPRI started to join hands in fostering a new disciplinary area of integrated disaster risk management. That year, IIASA and DPRI agreed to initiate the IIASA–DPRI Integrated Disaster Risk Management Forum Series. Eight annual forums were held under this initiative, helping to build a scholarly network that eventually evolved into the formation of the IDRiM Society in 2009.

These activities, which were designed to be cross-disciplinary and international, have seen synergistic developments. Japan's accumulated knowledge, led by DPRI, became merged with IIASA's extensive expertise and became connected with inputs from the USA, the UK, other parts of Europe, Asia, and other countries and regions.

Major Research Contributions

Among many, the following contributions merit mention:

Conceptual Models Developed and Shared for Integrated Disaster Risk Management Okada (2012) proposed systematic conceptual models for understanding the Machizukuri (citizen-led community management) approach. Figure 1 illustrates the multilayer common spaces (an extension of the concept of infrastructure) for a city, region, or neighborhood community as a living body (Okada 2004). This conceptual model has been found to be useful to address multilayer issues of integrated disaster risk management at various scales. For example, in the context of this diagram, Machizukuri is more appropriately applied on a neighborhood community scale rather than on a wider scale, such as a city or region. Applied to a neighborhood community in the context of a five-storied pagoda model, it starts with the fifth layer (daily life), followed by the fourth (land use and built environment) and the third (infrastructure). By comparison, Toshikeikaku (urban planning) focuses mainly on the fourth and third layers. Another point of contrast is that Machizukuri requires citizen involvement to induce attitudinal or behavioral change, while this issue is not essential for Toshikeikaku.

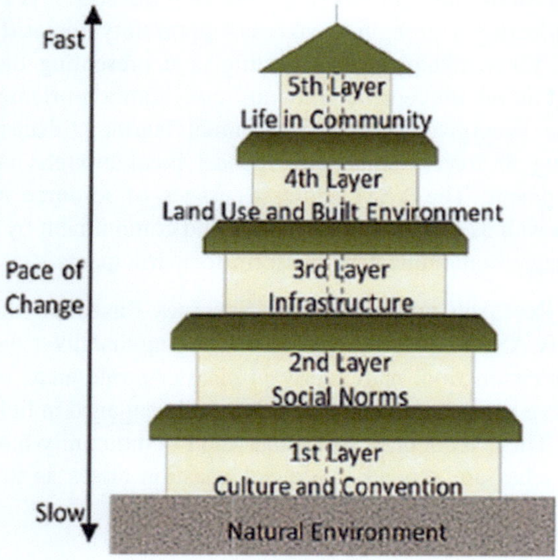

Fig. 1 Five-storied pagoda model (Source: Okada 2006)

Economic Modeling of Disaster Damage/Loss and Economic Resiliency Extensive research has been carried out by Tatano et al. (2004, 2007) and Tatano and Tsuchiya (2008) to model and analyze economic impacts of disruptions to lifelines and infrastructure systems caused by a large-scale disaster. For instance, simulating a hypothetical Tokai–Tonankai earthquake in Japan, a spatial computable general equilibrium (SCGE) model was constructed to integrate a transportation model that can estimate two types of interregional flows of freight movement and passenger trips. Kajitani and Tatano (2009) investigated a method for estimating the production capacity loss rate (PCLR) of industrial sectors damaged by a disaster to include resilience among manufacturing sectors. PCLR is fundamental information required to gain an understanding of economic losses caused by a disaster. In particular, this paper proposed a method of PCLR estimation that considered the two main causes of capacity losses as observed from past earthquake disasters, namely, damage to production facilities and disruption of lifeline systems. To achieve the quantitative estimation of PCLR, functional fragility curves for the relationship between production capacity, earthquake ground motion, and lifeline resilience factors for adjusting the impact of lifeline disruptions were adopted, while historical recovery curves were applied to damaged facilities.

Disaster Reduction-Oriented Community Workshop Methods The Cross-Road game developed by Yamori et al. (2007) proceeds as follows. During a game session, a group of five players read 10–20 episodes that are presented on cards one at a time. Each episode is derived from extensive focus group interviews of disaster veterans of the GHQ and describes a severe dilemma that the veterans of Kobe actually faced. Individual players are required to make an either/or decision (i.e., yes or no) between two conflicting alternatives in order to deal with the dilemma.

The Yonmenkaigi System Method (YSM) by Okada et al. (2013a, b) is a unique participatory decision- and action-taking workshop method. It is composed of four main steps: conducting a strength–weakness–opportunity–threat (SWOT) analysis, completing the Yonmenkaigi chart, debating, and presenting the group's action plan. The YSM is an implementation- and collaboration-oriented approach that incorporates the synergistic process of mutual learning, decision-making, and capacity building. It fosters small and modest breakthroughs and/or innovative strategy development. The YSM addresses issues of resource management and mobilization, as well as effective involvement and commitment by participants, and provides a strategic communication platform for participants.

Collaborative Research and Education Schemes Based on the Case Station-Field Campus (CASiFiCA) Scheme Acknowledging that diverse efforts have been made for disaster reduction, particularly in disaster-prone areas (countries), many professionals have been energetically and devotedly engaged in field work to reduce disaster risks. They recognize also that more community-based stakeholder-involved approaches are needed. A crucial question arises as to why we cannot

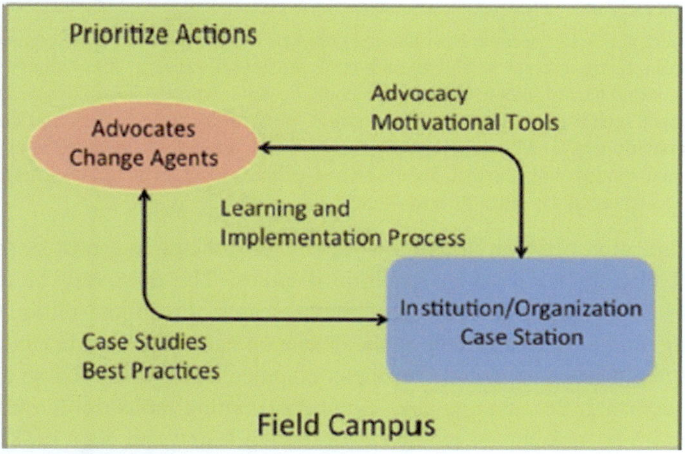

Fig. 2 Case Station-Field Campus scheme

conduct field work more creatively. One promising solution might be the CASiFiCA scheme originally proposed by Okada and Tatano (2008). As diagrammed in Fig. 2, the CASiFiCA scheme is characterized by a set of local case stations and field campuses and their globally networked linkages that are expected to operate synergistically to achieve the following objectives: promotion of IDRiM education at all levels, multilateral knowledge sharing and knowledge creation, and implementation of knowledge and gaining knowledge from implementation.

Europe

Integration via Regulation: European Union Experience

The integrated risk management of technological and natural hazard-triggered technological accidents (known as Natechs) has been a major theme addressed during the IIASA–DPRI Integrated Disaster Risk Management Forum Series since the first forum in 2001. In 2007 and 2008, the forum was hosted by the Major Accident Hazards Bureau at the Joint Research Centre of the European Commission in Italy, further strengthening the need for integration across natural and technological disaster risk management.

Integration was not (and, generally, still now *is not*) a self-evident concept when the first European Union Conference on Natural Risk and Civil Protection was launched in 1993, in Belgirate, Italy (Horlick-Jones et al. 1995). As the rapporteur-general wondered:

> Whilst one objective of the conference was to encourage dialogue between researchers and practitioners, it quickly became clear that the group structure was rather more complex than simply comprising natural scientists and civil protection experts. The 'tribes' present included natural hazard scientists, civil protection theorists—mostly social, behavioural and management scientists, industrial risk specialists, protection administrators and civil protection practitioners. The hazards and civil protection 'community' included a number of professional groups with distinct traditions and cultures. The term 'tribe' is used in an attempt to capture some sense of how strong is this divide.

Communication between the groups was rather difficult and most surprising for people not directly involved in scientific disputes. The discovery of the strong opposing views existing between different research directions within the same "hard" discipline (e.g., in seismology the debate on earthquake predictability) made even the agreement on an agenda for the conference challenging. These difficulties were unanticipated, because previous events concerning industrial hazards—organized in a similar manner on emergency planning (Gow and Kay 1988) and risk communication (Gow and Otway 1990)—found a rather cooperative atmosphere.

Despite the fact that the organization of the conference involved three directorate-generals of the European Commission (Research and Education, Environment, and Joint Research Center), natural hazards activities were not covered by an institutional legal basis. Also, at the time, there was no mutual assistance/compensation agreement in the case of a natural disaster, but only an initial exchange of experiences among emergency response services of EU member states. On the other hand, the existence of a sound regulatory process that obliged the different actors to be involved in the risk management framework was the reason for the successful cooperation in the latter mentioned events.

The new regulatory process for chemical accident prevention is an example. The process was reactive rather than anticipatory. It was triggered by a number of major accidents—e.g., the dioxin release at Seveso (Italy) in 1976 and the explosion at Flixborough (UK) in 1974. These had in common the features that local authorities did not know what chemicals were involved and in what quantities. They did not know enough about the processes to understand what chemicals/energy could be produced or released under accident conditions, and there was a general lack of planning for emergencies. Given this background, the first 1982 Seveso I Directive (82/501/EEC) was largely concerned with the generation and the control of an adequate and sufficient information flow among the different actors in the risk management process (Otway and Amendola 1989). This covered industrial activities that handle hazardous materials and introduced an integrated risk management scheme with identification of the actors and their obligations (control/licensing authorities — operators) or rights to know (the public). It requires that potential major accidents involving hazardous materials be identified, adequate safety measure be taken to prevent them, and on-site emergency plans be implemented. The competent authorities (CAs) have to control the adequacy of such measures and provide for external emergency plans. The public should be "actively" informed of the safety measures and how to behave in the event of an accident. The operator is required to report any major accident to the CAs, and the CAs have to notify the European

Commission, which keeps a register of accidents so that member states can benefit from this experience for the purposes of prevention of future accidents.

The Seveso I Directive was the background for further discussions at the international level, such as the Organisation for Economic Co-operation and Development (OECD) and the United Nations Economic Commission for Europe (UNECE), which resulted in further recommendations and conventions on trans-boundary effects related to major accidents (United Nations 1992).

Reacting to the tragedy in Bhopal, India and other issues identified during its implementation, the need for a revision was identified, particularly concerning the lack of provisions for land-use planning (De Marchi and Ravetz 1999), resulting in the Seveso II Directive (96/82/EC). It completed the transparency process, beginning with the obligation of disseminating information to the public on how to behave in case of an accident, and, in a relatively short time, changed the "secrecy" in most countries surrounded by chemical risks into unprecedented transparency (for the "evolutionary construction of a regulatory system" for an extensive discussion of all Seveso II requirements, see Amendola and Cassidy 1999). It established that the public should be consulted for land-use planning and emergency planning with respect to accident risks and therefore should be more directly involved in risk management decisions. Furthermore, the safety report and accident reporting systems became accessible by the public.

The Seveso II Directive focused much more on the socio-organizational aspects of the control policy:

- The concept of an industrial *establishment* was introduced, characterized by the presence of dangerous substances. The focus is on the interrelations among installations within such an establishment, especially those related to organization and management. Further, attention is given to situations liable to provoke so-called *domino effects* between neighboring establishments. This led to integrated assessments of industrial areas. Furthermore, it implicitly called for the analysis of external threats, such as natural hazards.
- The socio-organizational aspects of an establishment were strongly affected by the introduction of the obligation for a major accident prevention policy (MAPP), to be implemented by means of safety management systems (SMS) (Mitchison and Porter 1999). These provisions were introduced after the awareness that most of the major accidents of which the commission was notified over the years under the major accident reporting system (MARS) had root causes in faults of the management process (Drogaris 1993).
- The introduction of the obligation for a *land-use planning policy* with respect to major accident hazards has had important socio-organizational consequences, as a broader body of authorities, especially those dealing with local urban planning, are becoming involved in decisions about the compatibility of new development with respect to existing land use (Christou et al. 1999). This has been integrated with the requirement that the public shall be consulted in the decision-making process. This has also led to integration of planning policies with respect to other kinds of hazards, such as natural ones, assuring that

appropriate distances are kept between establishments, residential areas, and areas of particular "natural sensitivity."
- The provisions for *emergency planning* and *public information* have been reinforced, as the *safety report* becomes a public document, and the public must be consulted in the preparation of emergency plans.

The Seveso II Directive also approached management as a continuous process, because it did not limit the regulatory action to providing a license or a permit to operate. Instead it assigned the obligation to the operator to adopt management systems as a continuous process for feedback in the procedures relating to operating experience and managing the changes over time. Also, land-use planning addresses not only "siting" a new establishment but also considers the compatibility of major changes with the existing environment as well as the control of urbanization around an establishment. Furthermore, it promoted common efforts among authorities, operators, and risk analysts to improve the risk assessment procedures and achieve better risk governance processes (Amendola 2001).

As mentioned above, the Seveso II Directive called for the analysis of external hazards as part of the hazard assessment process. Both domino effects and land-use controls are of particular importance when addressing the risk reduction of chemical accidents triggered by external natural hazard events (Natechs). In fact domino effects may be more likely during natural disasters than during normal plant operation (Cruz et al. 2006; Lindell and Perry 1997). Their likelihood will depend on the proximity of vulnerable units containing hazardous substances, and the consequences will undoubtedly increase with the proximity of residential areas. The European Commission published guidelines to help member states fulfill the requirements of the Seveso II Directive (see Papadakis and Amendola 1997; Mitchison and Porter 1998; Christou and Porter 1999). However, the guidelines do not provide specific actions or methodologies that should be taken to prevent, mitigate, or respond to Natechs (Cruz et al. 2006).

In 2012, the European Commission published the Seveso III Directive, which amended and subsequently repealed the Seveso II Directive. The major changes included in the Seveso III Directive included strengthening of a number of areas such as public access to information and standards of inspections. Furthermore, the latest amendment now explicitly addresses Natech risks and requires that environmental hazards, such as floods and earthquakes, be routinely identified and evaluated in an industrial establishment's safety report (Krausmann 2016).

International Institute for Applied Systems Analysis (IIASA)

"Risk" has been part of IIASA's activity profile since the institute's foundation. This theme is critical, as the prospect of unintended consequences from technological, environmental, and social policies continues to stir intense debates that shape the future of societies across the world. Relying on probability calculations, risk became a theoretical focus designed to bolster a scientific, mathematically based approach toward uncertainty and risk management.

Early controversies in the 1970s and 1980s on nuclear power, liquid natural gas storage, and hazardous waste disposal—all early research topics at IIASA—made clear to the expert community, however, that probabilistic calculations of risk, although essential to the debates, are not sufficient to settle issues of public acceptance. In response, IIASA has pioneered research on risk perception (Otway and Thomas 1982), objective versus subjective assessments (Kunreuther and Linnerooth 1982), systemic cultural biases (Thompson 1990), and risk and fairness (Linnerooth-Bayer 1999).

As a critical part of this history, IIASA is widely recognized for its advances in stochastic and dynamic systems optimization (e.g., Ermoliev 1988), treating endogenous uncertainty and catastrophic risks in decision-making processes (reviewed in Amendola et al. 2013) and advancing statistical methods for probabilistic assessment (e.g., Pflug and Roemisch 2007). The hallmark of IIASA's risk research is the integration of these multiple strands of mathematical and social science research.

One important in-house model taking an integrated perspective in the RISK program at IIASA is the so-called Catastrophe Simulation (CatSim) Model, which focuses on the government and its fiscal risk in the face of natural disaster events. It is a mainstay of the program's methodological and policy research and was first developed to aid public officials in developing countries to assess catastrophic risks from natural hazards and analyze options to enhance their country's financial resiliency. The model takes a "systems approach" by integrating catastrophe risk modeling with financial and economic modeling. It enables users to explore the impact of traditional and novel financial instruments, including reinsurance and catastrophe bonds, in terms of the costs of reducing the risk of a financing gap. CatSim has proven useful in other contexts as well, e.g., for allocating climate adaptation and development funds to support disaster resilience in the most vulnerable countries. Based on the model framework, assessed exposure and financial vulnerability to extreme weather events on the global scale can be performed as well (Hochrainer-Stigler et al. 2014).

Beyond modeling, IIASA has pioneered the exploration of novel financing instruments to provide safety nets to vulnerable communities and governments facing climate risks (Linnerooth-Bayer and Amendola 2000). These instruments now feature prominently on the agendas of development organizations and NGOs, and they are also gaining attention in the climate change adaptation community (Linnerooth-Bayer and Hochrainer-Stigler 2015). In an early influential policy paper, IIASA scientists argued that donor-supported risk-transfer programs, some based on novel instruments, would leverage limited disaster-aid budgets and free recipient countries from depending on the vagaries of post-disaster assistance (Linnerooth-Bayer et al. 2005).

As a final mention, IIASA's contributions to integrated disaster risk management have included the design and implementation of new forms of bottom-up governance, most notably stakeholder processes which co-design policy options with experts and explicitly recognize large value differences.

The USA

Multidisciplinary Center for Earthquake Engineering Research

The National Center for Earthquake Engineering Research (NCEER) was established at the State University of New York at Buffalo in 1986, with funding from the US National Science Foundation (NSF), the state of New York, and industrial partners. NCEER's original vision focused on multidisciplinary research and education aimed at reducing earthquake losses. Although the Center's main priority was to support research in structural, civil, and geotechnical engineering, it also provided funding for research in the fields of economics, urban planning, regional science, and sociology. Despite NCEER's ambitious vision, much of the research conducted during the 10-year period of initial grant support remained discipline-specific, although with the passage of time there was greater integration across disciplines, particularly in areas such as earthquake loss estimation, which required collaborative approaches.

When NCEER leaders decided to enter a new competition for NSF funding in the mid-1990s, there was general agreement that investigators should step up their multidisciplinary collaborative efforts based on an understanding that earthquake risk reduction and risk management require contributions from a range of areas of expertise beyond traditional engineering fields. This was made explicit when the leadership decided to change the Center's name to the Multidisciplinary Center for Earthquake Engineering Research (MCEER). Participation in multidisciplinary teams was strongly encouraged as MCEER investigators increasingly tackled problems that were beyond the scope of individual disciplines. Experts in remote sensing and in structural engineering worked together on the development of building inventories and, later on, rapid post-earthquake damage assessment methods using remotely sensed data. Engineers, economists, and sociologists worked on improving earthquake loss estimation methods, focusing, for example, on estimating potential damage to urban lifeline systems as well as resulting direct and indirect economic losses. Collaborating teams developed earthquake recovery models and explored the economic, political, and institutional obstacles that stand in the way of adopting and implementing risk reduction policy. Researchers studied hospitals both as critical physical systems and as organizations. A multidisciplinary group consisting of engineers, policy experts, and decision scientists developed decision-support tools designed to help facility owners make informed choices about alternative seismic risk reduction measures.

In the late 1990s, another team of researchers from various fields began a series of projects focused on the conceptualization and measurement of earthquake (and general disaster) resilience. Recognizing that resilience itself is a multidisciplinary and even a transdisciplinary concept, researchers surveyed a wide range of studies in fields ranging from ecology to psychology, identified common concepts and indicators, and developed one of the first frameworks that applied the resilience concept to natural hazards. One early product resulting from that collaboration was the article "A Framework to Quantitatively Assess and Enhance the Seismic Resilience of

Communities" (Bruneau et al. 2003). Authors of that paper represented the fields of civil, geotechnical, and structural engineering, operations research, economic geography, decision science, and sociology.

These successful collaborations were the result of several factors. Research activities were problem focused, and the researchers involved recognized that the earthquake problem is multidimensional. Methodological tools such as geographic information systems were useful in bringing about integration across disciplines. The longevity of NCEER and MCEER was also important; long-term funding made it possible for investigators to engage with one another over prolonged periods. This also meant that over time, researchers came to better understand and appreciate the approaches and methods employed by their counterparts in other disciplines. Additionally, the intent of the funding source was a significant influence; NSF made it clear that it was looking for research that was capable of over-coming disciplinary silos.

A major example of integrated research at MCEER was the first New Madrid (Earthquake Zone) electricity lifeline case study (Shinozuka et al. 1998), which focused on the site of the largest earthquake to strike North America in its recorded history. The study team was composed of engineers, geographic information scientists, economists, regional scientists, planners, and sociologists. They addressed the complexity of the interaction of various systems in the Memphis Tennessee Metropolitan Area. This included the vulnerability of the lifeline network, business response to physical damage and production disruption, estimation of direct and indirect losses in the region and throughout the USA, and policy analysis and implementation. At the core of the research were models of economic, social, and spatial interdependence, such as input–output analysis, multisector mathematical programming, and social accounting matrices (all precursors of the now state-of-the-art approach of computable general equilibrium analysis). This research was performed around the same time as the development of FEMA's loss estimation software tool HAZUS (FEMA 1997, 2016), which was another example of an integrated assessment model (see also Whitman et al. 1997). The capabilities included in HAZUS had to be simplified in order to be incorporated into a decision-support system that could be used by a wide spectrum of emergency managers and analysts on a desktop PC. In contrast, the MCEER research was intended to advance the state of the art in improving the scope and accuracy of hazard loss estimation. As such, it proved valuable in future extensions and upgrades of HAZUS and informed other research and public and private decision-making. One of its major points was the prioritization of electricity service restoration according to various societal objectives such as minimizing lost production and employment. As one of the study authors noted: "Not taking advantage of such opportunities results in an outcome as devastating as if the earthquake actually toppled the buildings in which the lost production would've originated" (p. xvii).

MCEER was directed by Masanobu Shinozuka, George Lee and Michel Bruneau. Researchers who contributed to the integration of various disciplines under its umbrella, in addition to the directors, included Barclay Jones, Kathleen Tierney, Tom O'Rourke, Bill Petak, Charles Scawthorn, Detlof von Winterfeldt, Stephanie Chang, Ron Eguchi, and Adam Rose. Two sister centers of MCEER were established with NSF Funding in the mid-1990s: the Pacific Earthquake Engineering Center (PEER), headquartered at the University of California, Berkeley, with a focus on performance-based engineering; and the Mid-American Earthquake Center (MAE), headquartered at the University of Illinois, Urbana, with a focus on a multi-hazard approach to engineering.

Natural Hazards Center

The Natural Hazards Research and Applications Information Center at the University of Colorado Boulder—now called the Natural Hazards Center (NHC)—was founded in 1976 by Gilbert F. White, a geographer, and J. Eugene Haas, a sociologist. Center activities were built upon the foundation that White and his collaborators from many disciplines had already established, as outlined in the books *Natural Hazards: Local, National, and Global* (White 1976) and *Assessment of Research on Natural Hazards* (White and Haas 1975). In the *Assessment*, White and Haas argued that efforts to prevent and reduce disaster losses relied far too much on technological approaches, without taking into account research in the social sciences. Their position was that such research could offer important insights into societal responses to hazards and disasters while also shedding light on whether technological approaches aimed at reducing losses were likely to produce their intended outcomes. Early research assessments focused on "adjustments" to hazards that communities and societies can adopt either singly or in combination: relief and rehabilitation, insurance, warning systems, technological adjustments such as protective works, and land-use management. In the view of the founders, a key task for researchers was to better understand the conditions under which particular adjustments would be adopted and their subsequent impact on disaster losses. Early in its history, the NHC produced its own series of books, monographs, and special reports, many of which focused on findings from US National Science Foundation-sponsored research carried out by investigators in the social, economic, and policy sciences. That practice was discontinued as specialized journals began to proliferate and an increasing number of academic and commercial publishers began to show an interest in publishing research monographs and textbooks in the disaster field.

From its inception, the NHC has had a dual mission. First, it serves as a clearinghouse and information provider for social science research on hazard mitigation, preparedness, response, and recovery, again with an emphasis on alternative adjustments to hazards. The idea of an information clearinghouse arose out of recognition of the difficulties associated with getting research applied in real-world settings. Clearinghouse activities include the production and distribution of the NHC newsletter, the *Natural Hazards Observer,* library and information services, and the annual NHC workshop, which has grown over the years. From the beginning, the annual workshop was designed to bridge communication gaps among researchers and graduate students from a variety of physical, social science, and engineering disciplines, government decision-makers, and emergency management practitioners. The NHC also administers a small-grant quick-response research program that enables researchers and students to go into the field immediately following disasters and then publishes the results of those studies. Second, NHC faculty and graduate students conduct their own research, with support from the National Science Foundation and other sponsors.

Both the activities associated with the production of the original *Assessment* and subsequent center activities involved the training of young researchers from a variety of social science disciplines. The first generation of center graduate trainees included well-known researchers such as Harold Cochrane (economics); Eve Gruntfest and John Sorensen (geography); Dennis Mileti, Robert Bolin, and Patricia Bolton (sociology); and Michael Lindell (psychology).

During the 1990s, the NHC conducted the second assessment of research on natural hazards under the leadership of director Dennis Mileti. The second assessment, which involved contributions from approximately 120 researchers, students, agency personnel, and other public officials, resulted in five books and numerous published articles and reports, again reflecting a range of social science perspectives (e.g., Mileti 1999). Like its predecessor, the second assessment provided training for another generation of researchers.

Since the early 2000s, the NHC has been increasingly involved in multidisciplinary research projects. Examples include collaborations with computer scientists and other social scientists on new technologies for emergency management, with economists on post-disaster business and economic resilience, with researchers from the National Center for Atmospheric Research on warning systems, with investigators from a number of social science disciplines on homeland security-related issues, with engineering researchers on recovery from the 2004 Indian Ocean tsunami, and with engineers, earth scientists, and policy scientists on the problem of induced earthquakes.

The NHC has served under the able directions of its founders and successor directors geographer William Riebsame (now William Travis), sociologists Dennis Mileti and Kathleen Tierney, and, beginning in January 2017, sociologist Lori Peek.

Center for Risk and Economic Analysis of Terrorism Events (CREATE)

Soon after the September 11, 2001, terrorist attacks in the USA, the nation's National Academy of Sciences performed an assessment of how the scientific community, broadly defined, could contribute to reducing the terrorist threat. One of their recommendations was to establish university centers of excellence (COEs) in research and teaching. The first of these was the Center for Risk and Economic Analysis of Terrorism Events (CREATE), established in 2004 and headquartered at the University of Southern California but being a geographically distributed entity with more than a dozen affiliates at other universities and research organizations throughout the USA and some overseas. These faculty affiliates came from the

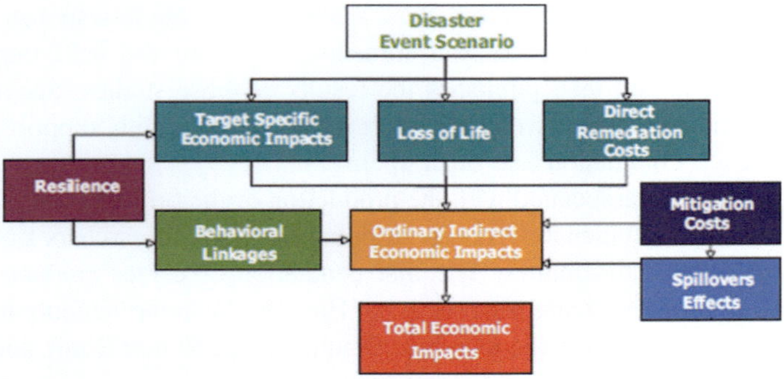

Fig. 3 CREATE economic consequence analysis framework

disciplines of decision analysis, risk analysis, psychology, economics, business, regional science, planning, operations research, public policy, public administration, public health, computer science, and communications. Founding directors were Randolph Hall and Detlof von Winterfeldt; subsequent directors were Stephen Hora and Ali Abbas, with von Winterfeldt returning after serving as director of IIASA.

Despite the restrictive nature of its title, CREATE was intended to be an "all hazards" center, although research in areas other than terrorism has been in the minority. CREATE was initially based on three themes: risk assessment, economic consequence analysis (and related topics in economics), and risk management. Risk communication was later inserted into the base of the framework. Much of the research has been multidisciplinary and some of it interdisciplinary.

One of the major interdisciplinary contributions was the development of a comprehensive framework for economic consequence analysis (ECA), as depicted in Fig. 3. This framework expanded ordinary economic impact analysis and hazard loss estimation substantially, first, by incorporating resilience. Building on his research at MCEER, Rose refined the concept of economic resilience into its static and dynamic versions, which are analyzed in the context of business interruption (BI), and focused the research on the demand, or customer, side, in terms of how businesses, households, and government agencies utilize remaining resources more efficiently and recover more quickly (see, e.g., Rose 2009 and this volume in the IDRiM Book Series). CREATE researchers performed many case studies using the operational metric that resilience effectiveness of any given strategy was equal to the averted BI as a proportion of the total potential BI in the absence of implementing the strategy. A major example was the finding that 72 % of the potential BI losses stemming from the destruction of the World Trade Center were averted by the rapid relocation of its business and government tenants (Rose et al. 2009).

Subsequent research has established the basis of an economic resilience index based on actionable variables (Rose and Krausmann 2013).

Another innovation was to incorporate "behavioral linkages," primarily off-site, post-disaster responses caused by such phenomena as the social amplification of risk and stigma effects. Many of these reactions are related to fear, as exemplified by the large BI following 9/11 from the decline of airline travel and related tourism (von Winterfeldt et al. 2006; Rose et al. 2009). A more in-depth and integrated analysis was undertaken to examine the BI losses from a simulated dirty bomb attack on the Los Angeles Financial District (Giesecke et al. 2012). This study examined the costs of potential wage and investor rate of return premia and customer discounts needed to attract people back to the targeted areas and inserted these costs in the state-of-the-art tool of economic consequence analysis—computable general equilibrium (CGE) modeling. The study results indicated that behavioral effects were 15 times larger than the ordinary direct and indirect economic impacts typically measured.

More recently, the framework has been "transitioned" to a user-friendly software tool known as E-CAT (Rose et al. 2017—a forthcoming volume in the IDRiM Book Series). A further extension of ECA on a parallel track to enhance the US government's terrorism risk assessment capability is being completed by Dixon and Rimmer (2016).

Other examples of interdisciplinary research at CREATE include work on adaptive adversaries, risk perceptions, risk messaging, and the value of information in risk management. This includes numerous case studies for academic and policy advising purposes that have been undertaken by CREATE researchers. One set of these has been the collaborative efforts between CREATE and the US Geological Survey (USGS) on analyzing disaster scenarios, such as a catastrophic earthquake, severe winter storm, tsunami, and massive cyber-disruption (see, e.g., Porter et al. 2011).

CREATE is one of a dozen COEs, with others involved in interdisciplinary research being the Consortium for the Study of Terrorism and Responses to Terrorism (START) and the Coastal Hazards Center. The centers have involved major researchers in the USA on both terrorism and natural hazards, such as Dennis Mileti, Kathleen Tierney, Susan Cutter, and Gavin Smith. An example of pioneering research is that on community resilience by Norris et al. (2008).

Low-Income Countries

It is difficult to pinpoint the beginning of academic research on natural hazards and disasters in low-income countries. The humanitarian system has deep historical roots, but the emergence of a humanitarian knowledge community is more recent and began to accelerate in the 1970s (Davey et al. 2013: 29). The 1970s and 1980s saw significant attention given to food emergencies and famine (Comité d'Information Sahel 1973; Sen 1981) and also to floods and cyclone impacts

(White 1976). The rapid growth of academic research in the 1970s and 1980s was arguably driven by the greater visibility and political saliency of disasters such as the famines in the West African Sahel and Ethiopia, huge loss of life in Bangladesh due to cyclones, and deadly earthquakes in Guatemala and China (Kent 1983; Wisner and Gaillard 2009). However, it was only in what the British call "development studies" that disaster vulnerability became a core concern during this early period, with, for instance, Chamber's introduction of the concept of vulnerability in the context of "integrated rural poverty" (1983) and theme issues of the *Bulletin of the Institute of Development Studies* devoted to problems of seasonality and to food security and the environment (Lipton 1986; Leach and Davies 1991). The international, interdisciplinary journal *Disasters* was launched in 1976. Geographers, political economists, anthropologists, students of international relations, and community health specialists were among the early contributors. Epidemiologists and other public health researchers were active in defining disasters as a new focus of research at about the same time (de Ville de Goyet 1976); however, they worked alone or in small groups. The large academic center devoted to interdisciplinary, integrated approaches to understanding and managing disasters in low-income countries is a more recent development.

National Interdisciplinary Centers in the Global North

In the early twenty-first century, dedicated research centers now exist whose staff and collaborators span disciplines from the earth science and geoinformatics, social work, engineering, and public health to psychology, economics, sociology, politics, and geography, among others. Their approach is generally applied to and focused on the policy and practice of management of disaster prevention and risk reduction, warning, response and relief, and recovery. Two examples are the IRDR at University College London and IHRR at Durham University.

The Institute for Risk and Disaster Reduction (IRDR https://www.ucl.ac.uk/rdr) at University College London draws from a wide range of the University's institutes and departments, including the Institute for Global Health, Development Planning Unit in the Bartlett School of Architecture, Faculty of Engineering Sciences, the Leonard Cheshire Disability and Inclusive Development Centre, and departments of earth science and psychology, among many others. IRDR affiliates conduct research on the public perception of risk and how diverse societies deal with disaster, understanding health risks and pandemics, the study of extreme weather and the climate forcing of geological hazards, innovative design and construction, planning and design codes, and issues of resilience and recovery. One UCL partner with IRDR, the UCL Hazard Centre, has placed Ph.D. student researchers in nongovernmental development organizations (NGOs) in order to enhance NGO effectiveness (https://www.ucl.ac.uk/hazardcentre/ngo).

The Institute of Hazard, Risk and Resilience (IHRR https://www.dur.ac.uk/ihrr/) covers a similar range of research topics and also engages staff and research students across many disciplines at the University of Durham. IHRR plays a central role in the Earthquakes Without Frontiers research program in a number of countries in the Alpine–Himalayan Belt. This work involves earth scientists, social scientists, a historian, and a professor of social work and seeks to understand secondary earthquake hazards such as landslides, as well as risk governance and perception of earthquake risks by stakeholders at a number of scales (http://ewf.nerc.ac.uk/). IHRR researchers are also investigating such health aspects of disaster management as the effectiveness of respiratory protection during volcanic eruptions and economic questions such as how well small and medium enterprises recover from flooding.

International Centers

Because the elimination of poverty and promotion of security for people from food shortage, disease, and natural hazards are among the mandates of a number of UN organizations and international organizations, it is not surprising that research on integrated disaster risk reduction and management also takes place in these institutional homes. The World Bank and United Nations Development Programme (UNDP) are keenly aware of risk and are active on issues of human security (World Bank 2014; UNDP 2014). The World Health Organization (WHO) and the World Food Programme (WFP) also commission and conduct research on the early warning and management of epidemics and food emergencies, respectively (WHO 2016; WFP 2016). The Intergovernmental Panel on Climate Change (IPCC) has addressed the impacts of climate change on poor people in poor countries, particularly in its major report on climate-related disasters (IPCC 2012).

Also at the international scale, a good deal of the work of IIASA has been important in shaping policy and practice of risk management in low-income countries, for example, in the area of disaster insurance. The Center for Research on the Epidemiology of Disasters at the Catholic University of Louvain (CRED) in Belgium has evolved from a collector and repository of disaster data into a multifunctional academic institution that also produces occasional reports of relevance to integrated disaster risk management. One example is its 2016 report on poverty and disaster deaths (CRED 2016).

The International Council for Science has launched an initiative on the integrated study of disaster risk (http://www.irdrinternational.org/). Based in Beijing, China, the program of Integrated Research on Disaster Risk (IRDR) is active worldwide, especially in the Global South. It encourages young scientists, and it is currently engaged in an international assessment of integrated research on disaster that may lead to the IRDR's becoming the hub of a community of practice for such work. Its other research areas include knowledge sharing on the assessment of disaster loss and of the factors involved in the ways that people make decisions regarding disaster risk. In all of these functions, the emphasis is on serving a networking and facilitating function among researchers.

Another major program at IRDR has been to develop a framework for the forensic analysis of disasters called Forin (IRDR 2015). It seeks to focus researchers' attention on the root causes of disaster that go beyond the physical triggering phenomena and simple human exposure. Forin is grounded in a theory of social construction of disaster risk (Wisner et al. 2004, 2016; Tierney 2014). While keenly aware of physical and biological processes that manifest as hazards, Forin focuses on the process of development itself as a locus of risk creation (Oliver-Smith et al. 2016).

The forensic approach of the IRDR's Forin framework is not unusual. For many researchers who come to disaster risk from a background of work on poverty and marginalization in low-income countries, disaster is understood as a manifestation of failed or distorted development (Lavell et al. 2012) and the accumulation of risk in everyday life (Bull-Kamanga et al. 2003). Data collected beginning in the early 1970s shows that marginalized and excluded social groups in formerly colonized and other low-income countries are more severely impacted by natural hazards (Wisner et al. 2004). Women die in greater numbers in floods and coastal storms. Small farmers and fishers end up losing their land and boats to more wealthy neighbors and money lenders and find it more difficult to reestablish viable livelihoods.

The perspective of research grounded in daily realities of the urban and rural poor has also revealed that local knowledge and ways of adapting to hazards have been overlooked by planners and managers. In the last two decades, there has been much research on how local knowledge of hazardous environments can be brought together with outside specialist knowledge (Wisner 1995, 2010, 2016). The concept and practice of community-based disaster risk management (CBDM) or risk reduction (CBDR) have become common among both academic researchers and a large number of nongovernmental organizations, and collaboration between civil society and academia has begun in this domain (Wisner et al. 2008; Kelman and Mercer 2014).

National and Regional Centers in the Global South

Interdisciplinary research is also being conducted by institutions within low- and medium-income countries themselves. In the Americas, the network of researchers known as La Red was a pioneer (http://www.desenredando.org/). Created in 1992, La Red has a relationship with FLACSO, the graduate faculty of social sciences shared by ten Latin American countries. La Red publishes a journal, *Sociedad y Desastres* (http://www.desenredando.org/public/revistas/dys/), suspended for a time, but now relaunched, and has incubated some of the world's most innovative work on participatory action research for disaster reduction and on deep analysis of the links between development and disaster. Many of these innovations, while originally focused on the region and published in Spanish, have taken on an international role in shaping how disaster is understood and measured. A disaster monitoring and inventory tool known as DesInventar (http://www.desinventar.org/)

was created by associates of La Red. It makes use of sub-national media and civil society sources to catalogue small- and medium-scale hazard events that have been shown to have a major impact on livelihoods and human security. Since its earliest application in Colombia, it is now used in many parts of the world.

In South Africa, Stellenbosch University and North-West University have interdisciplinary centers devoted to disaster risk management. At Stellenbosch, the Research Alliance for Disaster Risk Reduction (RADAR) began in 2013 to build on 17 years of research and networking on the continent when the director was based at Cape Town University. A large body of work on urban disaster risks such as shack fires and risk management in South Africa has resulted, as well as work on flooding. In addition, Peri Peri University is coordinated from a base in RADAR (http://www.riskreductionafrica.org/partners-and-programmes/stellenbosch-university-stellen-bosch-south-africa/.). Peri Peri U is a network of 11 universities in sub-Saharan Africa that share knowledge on disaster-focused pedagogy and research methods. North-West University is home to the African Centre for Disaster Studies (ACDS http://acds.co.za/). Established in 2002, ACDS conducts research on disaster risk governance, gender and disasters, water-related risks, and climate change. It is also home to a peer-reviewed, open-access journal, *Jàmbá: Journal of Disaster Risk Studies* (http://www.jamba.org.za/index.php/jamba).

In South Asia, a group of researchers pulled from civil society, journalism, and academia produces the occasional *South Asia Disaster Report* (e.g., Practical Action 2010) coordinated by the NGO called Duryog Nivaran and facilitated over the years by the INGO, Practical Action.

Many of the participants in these various research efforts in the Asia-Pacific region, the Middle East, Africa, Latin America, and the Caribbean have collaborated over the years with research into local, lived realities of disaster risk and risk reduction. The Global Network of Civil Society Organisations for Disaster Reduction (GNDR www.gndr.org) has in this way been able to mount large surveys that involved 800 civil society organizations in 129 countries, tapping the knowledge of more than 85,000 respondents in its Views from the Frontline series (http://www.gndr.org/programmes/views-from-the-frontline/vfl-2013.html), as well as even more detailed studies of local risk perception and action in its Frontline and Action at the Frontline series (Gibson and Wisner 2016).

Summary

The examples provided above are not exhaustive. Groups of researchers in many universities, civil society organizations, and government departments in low- and medium-income countries carry out work on disaster risk, albeit some of it more and some less integrated and interdisciplinary, given differences in the history of relations among academia, news media, and government and differences in bureaucratic flexibility within higher education and government. The important takeaways from this brief overview are that:

- A vital and growing focus on disaster risk in low- and medium-income countries has emerged
- A consensus is growing that disaster risk in such countries is to a great degree a manifestation of failed development
- The applied focus on practice and policy leads such research toward an integrated management approach
- Systemic changes in governance and in the relations among academia, civil society (including the media), and government are necessary if research on integrated risk management is to flourish in low- and medium-income countries themselves, and elsewhere in the Global South, as opposed to relying primarily on work within rich-country institutions and international organizations in the Global North

Other Contributions

The brief summaries of research contributions on integrated disaster risk management presented above are not all-inclusive. They focus to a great extent on work performed through major research institutions. As such, they omit contributions by several who have contributed to the IDRiM cause before the formation of the organization and since. Some examples are noted below.

The interrelationship between disasters and development was given a significant boost by the establishment of a program in disaster and development studies at Northumbria University (UK) in 2000 (see also the Department of Geography/Disaster and Development Network, DDN). This also co-emerged with integration of more specialized fields such as health and well-being-centered disaster risk reduction and communities and resilience, all of which are based on integrated approaches. Early work by Andrew Collins and others focused specifically on infectious disease risk management, bringing together microbial ecology, socio-behavioral, and contextual analyses to identify best-integrated risk management practices in Mozambique and Bangladesh (see http://www.ukcds.org.uk/the-global-impact-of-uk-research/communities-against-disasters). A broader set of universities are involved in the UK Alliance for Disaster Research (UKADR) (www.ukadr.org).

In Austria, BOKU University has a long tradition in the research of water resources, including current involvement in the South East Europe (SEE) project CC-WARE (Mitigating Vulnerability of Water Resources Under Climate Change). It is led by the forest section of the Austrian Federal Ministry of Agriculture, Forestry, Environment and Water Management and includes 17 partners from 10 countries. The main objective of CC-WARE is the development of an integrated transnational strategy for water protection and mitigating water resources vulnerability as a basis for the implementation of national and regional action plans (http://www.ccware.eu/). See also Löschner et al. (2016).

DPRI, with funding from the government of Japan under its GCOE Human Security Engineering (HSE) initiative, promoted field-based research projects on disaster risk management in Asian megacities. The Mumbai project, 2009–2013, focusing on vulnerable hot-spot communities, was established with the objective of evolving scientific methodology on participatory grassroot-level disaster risk management. The project, a first of its kind in India and one among a few globally, was undertaken in collaboration with the Mumbai city government (MCGM); School of Planning and Architecture, New Delhi; the Tata Institute of Social Science; IIT Bombay; and JJ School of Architecture, Mumbai. One outcome is a breakthrough in process methodology that empowered the two hot-spot poor communities to play the lead role in what is known as community-based disaster risk management (CBDRM). IDRiM founding member Bijay Anand Misra served as the senior adviser and coordinator of the project (see Misra 2013).

IDRiM member Manas Chatterji has overlapped research on integrated disaster risk management with work on conflict management and peace science (see, e.g., Chatterji et al. 2012).

Several research centers working on aspects of integrated disaster risk management operate in Iran, such as the International Institute of Earthquake Engineering and Seismology, under the founding and long-term leadership of Professor Mohsen Ghafory-Ashtiany, who also serves as the Chairman of the SP Insurance Risk Management Institute.

As one major example of research in China, in 2011, the Risk Governance Group of the Chinese National Committee on International Dimensions Programme on Global Environmental Change (CNC-IHDP) launched its Integrated Risk Governance (IHDP-IRG) Project. As a ten-year international cooperative research effort, its mission is to improve the governance of new risks that exceed current human coping capacities by focusing on the transitions in and out of the occurrence of relevant risks in the global climate changes. Under this project Beijing Normal University, with the leadership of Peijun Shi and others, has led comprehensive scientific research that included the several case studies, a community risk governance model, and a proposed paradigm of catastrophe risk governance in China. See, e.g., Shi et al. (2013) for a comparative study of the Wenchuan Earthquake and Tangshan Earthquake, centering on hazard, exposure, disaster impacts and losses, disaster rescue and relief, and recovery and reconstruction.

Limitations of space restrict us from mentioning all those working on the topic of resilience, but, in addition to the people and organizations mentioned above, we note the following whose research is in the spirit of integrated disaster risk management: Erica Seville, co-Leader of the Resilient Organisations community in New Zealand, Stephane Hallegatte of the World Bank, and Swenja Surminski of the Overseas Development Institute.

Conclusion

Further efforts needed in the future to advance integrated disaster risk management include:

- Extending research perspectives and constructing new conceptual models
- Developing new methodologies
- Exploring yet uncovered and newly emerging phenomena and issues
- Engaging in proactive field studies in regions that face high disaster risks, but, where investigations have not yet been undertaken, performing field studies that incorporate research advances in disaster-stricken regions

Obviously, the above approaches are rather interdependent, and thus integrated disaster risk management is best promoted by combining them. For instance, emerging mega-disasters, which are caused by an extraordinary natural hazard taking place in highly interconnected societies, may require a combination of both the second and third points above, such as mega-disaster governance based in part on mathematical models of systemic risks. Also, long-range planning for societal implementation of integrated disaster risk management inevitably requires encompassing most of the above approaches.

The IDRiM Book Series as a whole intends to cover most of the aforementioned new research challenges.

Nishinomiya, Japan — Norio Okada
Milan, Italy — Aniello Amendola
Laxenburg, Austria — Joanne Bayer
Uji, Japan — Ana Maria Cruz
Laxenburg, Austria — Stefan Hochrainer
Los Angeles, CA, USA — Adam Rose
Boulder, CO, USA — Kathleen Tierney
Oberlin, OH, USA — Ben Wisner

References

Amendola A (2002) Recent paradigms for risk informed decision-making. Saf Sci 40(1–4):17–30

Amendola A, Cassidy K (eds) (1999) Special issue of the J Hazard Mater 65(1–2)

Amendola A, Ermolieva T, Linnerooth-Bayer J, Mechler R (eds) (2013) Integrated catastrophe risk modeling: supporting policy processes. Springer, Dordrecht

Bruneau M, Chang S, Eguchi R, Lee G, O'Rourke T, Reinhorn A, Shinozuka M, Tierney K, Wallace W, von Winterfeldt D (2003) A framework to quantitatively assess and enhance seismic resilience of communities. Earthq Spectra 19:733–752

Bull-Kamanga L, Diagne K, Lavell A, Leon E, Lerise F, MacGregor H, Maskrey A, Meshack M, Pelling M, Reid H, Sattertwaite D, Songsore J, Westgate K, Yitambe A (2003) From everyday hazards to disasters: the accumulation of risk in urban areas. Environ Urban 15(1):193–204

Chatterji M (ed) (2013) Cooperation for a peaceful and sustainable world, Part 1. Emerald Publishing, Bradford

Christou MD, Porter S (eds) (1999) Guidance on land use planning as required by Council Directive 96/82/EC (Seveso II). EC/JRC, Report EUR 18695 EN, Luxembourg.

Christou M, Amendola A, Smeder M (1999) The control of major accident hazards: the land-use planning issue. J Hazard Mater 65(1–2):151–178

Comité d'Information Sahel (1973) Qui se nourrit de la famine en Afrique? Maspero, Paris

CRED (Center for the Epidemiology of Disaster) (2016) Poverty and disaster deaths. Louvain, Belgium. http://cred.be/sites/default/files/CRED_Disaster_Mortality.pdf

Cruz AM, Steinberg LJ, Vetere-Arellano AL (2006) Emerging issues for Natech disaster risk management in Europe. J Risk Res 9(5):1–19

Davey E, Bolton J, Foley M (2013) A history of the humanitarian system. Humanitarian Working Group, Overseas Development Institute, London

De Marchi B, Ravetz JR (1999) Risk management and governance: post-normal approach. Futures 31(7):743–757

de Ville de Goyet C (1976) L' epidémiologie des désastres: Une science naissante au service des pays en voie de développement. Louvain Médical: Revue du Secteur des Sciences de la Santé de l'Université Catholique de Louvain 95:145–155 & 185–195

Dixon P, Rimmer M (2016) Economic implications of terrorism risk assessment scenarios: a dynamic computable general equilibrium analysis. Report to the U.S. Department of Homeland Security, forthcoming

Drogaris G (1993) Learning from major accidents involving dangerous substances. Saf Sci 16 (2):89–113

Ermoliev Y (ed) (1988) Numerical techniques for stochastic optimization. Springer, New York

Federal Emergency Management Agency (1997) HAZUS. Washington, DC Federal Emergency Management Agency (2016) HAZUS-MH. Washington, DC

Ghafory-Ashtiany M (2014) Earthquake risk of and risk reduction capacity building. In: Ismail-Zadeh A (ed) Extreme natural hazards, disaster risks and societal implications. Cambridge University Press, Cambridge

Gibson T, Wisner B (2016) Lets talk about you …: opening space for local experience, action and learning in disaster risk reduction. Disaster Prev Manag: An International Journal 25(5):664–684

Giesecke JA, Burns WJ, Barrett A, Bayrak E, Rose A, Slovic P, Suher M (2012) Assessment of the regional economic impacts of catastrophic events: CGE analysis of resource loss and behavioral effects of an RDD attack scenario. Risk Anal 32(4):583–600

Gow HBF, Kay RW (eds) (1988) Emergency planning for industrial hazards. Elsevier Applied Science, London and New York

Gow HBF, Otway H (eds) (1990) Communicating with the public about major accident hazards. Elsevier Applied Science, London and New York

Hochrainer-Stigler S, Mechler R, Pflug GC, Williges K (2014) Funding public adaptation to climate-related disasters. Estimates for a global fund. Global Environ Chang 25:87–96

Horlick-Jones T, Amendola A, Casale R (eds) (1995) Natural risk and civil protection. E&FN Spon, London

IPCC (Intergovernmental Panel on Climate Change) (2012) Special report on managing the risks of extreme events and disasters to advance climate change adaptation. Geneva. http://ipcc-wg2.gov/SREX/report/

IRDR (Integrated Research on Disaster Risk) (2015) The Forin project. Beijing, China. www.irdrinternational.org/wp…/03/FORIN-Case-Studies-Booklet-WEB-13-MB.pdf

Kajitani Y, Tatano H (2009) Estimation of lifeline resilience factors based on surveys of Japanese industries. Earthquake Spectra 25(4):755–776

Kent R (1983) Reflecting upon a decade of disasters: the evolving response of the international community. Int Aff 59(4):693–711

Krausmann E, Cruz AM, Salzano E (eds) (2016) Natech risk assessment and management: reducing the risk of natural-hazard impact on hazardous installations. Elsevier

Kunreuther H, Linnerooth-Bayer J (1983) Risk analysis and decision processes: the siting of liquefied energy gas facilities in four countries. Springer Verlag, Berlin

Lavell A, Gaillard JC, Wisner B, Saunders W, van Niekerk D (2012) National planning and disaster. In: Wisner B, Gaillard JC, Kelman I (eds) The Routledge handbook of hazards and disaster risk reduction. Routledge, London, p 617–628

Leach M, Davies R (eds) (1991) Food security and the environment. Theme issue, IDS Bulletin 22 (3)

Lindell MK, Perry RW (1997) Hazardous materials releases in the Northridge earthquake: implications for seismic risk assessment. Risk Anal 17(2):147–156

Linnerooth-Bayer J (1999) Climate change and multiple views of fairness. In: Toth FL (ed) Fair weather? Equity concerns in climate change. Earthscan Publications Ltd., London, p 44–65

Linnerooth-Bayer J, Amendola A (2000) Global change, natural disasters and loss-sharing: issues of efficiency and equity. Geneva Pap Risk Insur Issues Pract 25:203

Linnerooth-Bayer J, Hochrainer-Stigler S (2015) Financial instruments for disaster risk management and climate change adaptation. Clim Chang 133(1):85–100

Lipton M (ed) (1986) Seasonality and ultrapoverty. Theme issue. IDS Bull 17(3):4–8

Löschner L, Herrnegger M, Apperl B, Senoner T, Seher W, Nachtnebel HP (2016) Flood risk, climate change and settlement development: a micro-scale assessment of Austrian municipalities. Reg Environ Change, forthcoming

Mileti D (1999) Disasters by design. Joseph Henry Press, Washington, DC

Misra, BA (2013) Disaster risk management: challenges and conflicts in 21st century governance, mega city Mumbai, in Disaster Risk Management Conflict and Cooperation, Concept Publishing Company

Mitchison N, Porter S (eds) (1999) Guidelines on a major accident prevention policy and safety management system, as required by Council Directive 96/82/EC (SEVESO II). EUR 18123 EN, JRC, Ispra

Norris F, Stevens S, Pfefferbaum V, Whyche K, Pfefferbaum R (2008) Community resilience as a metaphor, theory, set of capacities and strategy for disaster readiness. Am J Community Psychol 41:127–150

Okada N (1985) Putting disaster prevention into a perspective of risk management. Tsuchi to Bosai (Soils Management and Disaster Prevention), Japan Society for Civil Engineers

Okada N (2006) Perspective on integrated disaster risk management. In: Hagihara Y, Okada N, Tatano H (eds) Introduction to integrated disaster risk management. Kyoto University Academic Press, Kyoto, pp 9–54

Okada N (2012) Lessons learned from recent disasters in Japan, and implications for ASEAN countries. Keynote presentation at ASEAN-Japan meeting, Naha, 26 June 2013

Okada N, Tatano H (2008) Case Station-Field Campus (CASiFiCA): globally-networked, field-based research and education challenges for disaster reduction. Presented at IDRC DAVOS Conference 2008, Davos, Switzerland. https://idrc.info/fileadmin/user_upload/idrc/former_conferences/idrc2008/presentations/Program_forWebsite_V7.pdf#search=Davos+conference+Case+StationField+Campus+%28CASiFiCA%29+%3A+Globallynetworked%2C+Fieldbased+Research+and+Education+Challenges+for+Disaster+Reduction

Okada N, Fang L, Kilgour M (2013a) Community-based decision making in Japan. Group Decis Negot 22(1):45–52

Okada N, Na J, Fang L, Teratani A (2013b) The Yonmenkaigi system method: an implementation-oriented group decision support approach. Group Decis Negot 22(1):53–67

Oliver-Smith A, Alcántara-Ayala I, Burton I, Lavell A (2016) Forensic investigation of disasters: a conceptual framework and guide to research. Beijing. www.irdrinternational.org/wp-content/uploads/2016/01/FORIN-2-29022016.pdf

Otway H, Thomas K (1982) Reflections on risk perception and policy. Risk Anal 2:69–82

Papadakis GA, Amendola A (eds) (1997) Guidance on the preparation of a safety report to meet the requirements of Council Directive 96/82/EC (SEVESO II). EC/JRC, Report EUR 17690 EN, Luxembourg
Pflug G, Römisch W (2007) Modeling, measuring and managing risk. World Scientific, Singapore
Porter K, Jones L, Cox D, Goltz J, Hudnut K, Mileti D, Perry S et al. (2011) The ShakeOut scenario: a hypothetical Mw7. 8 earthquake on the southern San Andreas fault. Earthquake Spectra 27(2):239–261
Practical Action, Nivaran D (2010) South Asia disaster report 2010: changing climate, impeding risks, emerging perspectives. Colombo. http://practicalaction.org/disaster-report
Rose A (2009) Economic resilience to disasters. Community and Regional Resilience Institute (CARRI) Research Report Number 8
Rose A, Krausmann E (2013) An economic framework for the development of a resilience index for business recovery. Int J Disaster Risk Reduct 5:73–83
Rose A, Oladosu G, Lee B, Assay G (2009) The economic impacts of the September 11 terrorist attacks: a computable general equilibrium analysis. Peace Econ Peace Sci Publ Policy 15(2)
Rose A, Prager F, Chen Z, Chatterjee S, Wei D, Heatwole N, Warren E (2017) Economic consequence analysis of disasters: the E-CAT software tool. Springer, Tokyo, forthcoming
Sen A (1981) Poverty and famines. Oxford University Press, Oxford
Shi P, Jaeger C, Ye Q (2013) Integrated risk governance-science plan and case studies of large-scale disasters. Beijing Normal University Press/Springer, Heidelberg/Berlin
Shinozuka M, Rose A, Eguchi R (1998) Engineering and socioeconomic impacts of earthquakes: an analysis of electricity lifeline disruptions in the New Madrid Area. Multidisciplinary Center for Earthquake Engineering Research Buffalo, New York
Tatano H, Tsuchiya S (2008) A framework for economic loss estimation due to seismic transportation network disruption: a spatial computable general equilibrium approach. Nat Hazards 44(2):253–265
Tatano H et al (2004) Economic restoration after a catastrophic event: heterogeneous damage to infrastructure and capital and its effects on economic growth. J Nat Dis Sci 26(2):81–85
Tatano H, Tsuchiya S, Okada N (2007) Economic loss assessment due to railroad and highway disruptions. Econ Syst Res 19(2):147–162
Thompson M, Ellis R, Wildavsky A (1990) Cultural theory. Westview Press, Boulder
Tierney K (2014) The social roots of risk: producing disasters, promoting resilience. Stanford University Press, Stanford
UNDP (United Nations Development Programme) (2014) Human development report 2014: sustaining human progress: reducing vulnerabilities and building resilience. UNDP, New York
von Winterfeldt D, O'Sullivan TM (2006) Should we protect commercial airplanes against surface-to-air missile attacks by terrorists? Decis Anal 3(2):63–75
WFP (World Food Programme) (2016) Comprehensive food security and vulnerability analysis. https://www.wfp.org/food-security/assessments/comprehensive-food-security-vulnerability-analysis
White G (ed) (1976) Natural hazards: local, national, and global. Oxford University Press, New York
White G, Haas E (1975) Assessment of research on natural hazards. Cambridge, MIT Press
Whitman R et al (1997) Development of a national earthquake loss estimation methodology. Earthquake Spectra 13(4):643–661
WHO (World Health Organization) (2016) Early warning systems. http://www.who.int/csr/labepidemiology/projects/earlywarnsystem/en/
Wisner B (1995) Bridging "expert" and "local" knowledge for counter-disaster planning in urban South Africa. GeoJournal 37(3):337–348
Wisner B (2010) Climate change and cultural diversity. Int J Soc Sci 61(199):131–140
Wisner B (2016) Vulnerability as concept, model, metric and tool. In: Cutter S (ed) Oxford research encyclopedia of natural hazard science. Oxford University Press, New York. Online only: http://naturalhazardscience.oxfordre.com/view/https://doi.org/10.1093/acrefore/9780199389407.001.0001/acrefore-9780199389407-e-25

Wisner B, Gaillard JC (2009) An introduction to neglected disasters. Jàmbá: J Dis Risk Studs 2(3):151–158

Wisner B, Blaikie P, Cannon T, Davis I (2004) At risk: natural hazards, people's vulnerability and disaster, 2nd edn. Routledge, London

Wisner B, Haghebaert B, Schaerer M, Arnold M (2008) Community risk assessment toolkit.
ProVention Consortium, Geneva. Online only: http://www.proventionconsortium.net/?pageid=39

World Bank (2014) World development report 2014: risk and opportunity. World Bank, Washington, DC

Yamori K (2007) Disaster risk sense in Japan and gaming approach to risk communication. Int J Mass Emerg Dis 25(2):101–131

Introduction

This book is one volume in the Integrated Disaster Risk Management series and focuses on the theme of disaster risk communication. The foundation of this type of communication is disaster risk information, which is the knowledge needed to determine either the risk of a disaster itself or the damage it is expected to cause. Examples of disaster risk information include a weather advisory stating a 50% likelihood that an approaching hurricane will make landfall, a predicted maximum height of 20 meters for a tsunami accompanying a large earthquake expected to hit in the near future, or an estimate that this tsunami will kill approximately 10,000 people in the worst-case scenario.

It goes without saying that the aim of producing disaster risk information and communicating it to society is to reduce the damage caused by disasters. Predictions of the path and scale of hurricanes or the scale of tsunamis and their potential damage are intended to encourage efforts to reduce the impacts of these disasters, for instance, through early evacuation and advanced designation of evacuation sites. Put another way, this type of disaster risk information is communicated to society for the precise reason that it should lessen the impacts in comparison with situations where disasters catch communities by surprise. Indeed, disaster risk information has contributed to disaster mitigation in many cases.

However, we must not overlook the fact that in regions such as North America, Europe, and Japan, where the natural sciences and R&D based on them have advanced and the quality and quantity of disaster risk information have consequently exceeded a certain level, the opposite phenomenon is also occurring. Knowledge (or excessive knowledge) about the risk of disasters—that is, the effective communication of these risks to society—can itself cause harm.

This is what happened after the Great East Japan Earthquake in 2011, when the massive earthquake and tsunami damage led to heightened fear of a similar earthquake and tsunami hitting central or western Japan, where such disasters have occurred periodically throughout history, as a result of seismic activity in the Nankai Trough. In 2012, the government officially announced the results of a reanalysis of Nankai Trough earthquake and tsunami risk. The dire contents of this report predicted that just ten-some minutes after an earthquake hit, a tsunami measuring over 30 meters high and exceeding those of the 2011 disaster could wash over western Japan, killing up to 320,000 people (over 15 times the number who died in the disaster of 2011).

The societal impacts of this disaster risk information deserve consideration. The information did, of course, lead to some efforts aimed at reducing damage from a possible future disaster. Many new evacuation facilities were constructed, and disaster drills took place with greater frequency. On the other hand, a significant number of people reacted by giving up on the possibility of evacuation (in other words, of reacting in ways that could reduce the level of damage), telling themselves that it would be no use trying to escape, or that they lacked the physical strength to get away from such an overwhelming tsunami. This reality indicates that the communication of disaster risk information caused reactions that were unanticipated, and in fact completely the opposite of what was initially hoped.

The pitfalls of disaster risk communication take other forms as well. In many cases, disaster risk is communicated from individuals who already possess the information (typically a disaster expert or emergency preparedness manager at a municipal government or similar body) to those who do not (typically residents of the local community and nonexperts). However, this act of communication has the side effect of reinforcing, magnifying, and reproducing dichotomies between the teachers and the taught, or those who know and those who do not. This, in turn, leads to situations such as disaster drills in which experts and municipal employees participate actively and independently while the all-important community members either do not participate at all or passively follow the lead of the experts.

These observations reveal the importance not only of determining or predicting disaster risks and producing reliable disaster risk information, but also the even greater importance of considering the specific effects that occur when that information is communicated to society, particularly the unintended side effects and reactions it can cause. In other words, there is a need for research and on-the-ground initiatives related to the communication of disaster risk itself. It is particularly crucial to develop communication styles and techniques that deter unwanted side effects and reactions, and instead induce constructive attitudes and actions that will reduce the damage from disasters. Above all, the development of more participatory communication techniques is essential.

This book takes on these challenges. Rather than focus solely on content when studying and practicing communication, we must turn our attention to communication style (that is, who communicates the information, and in what manner). In order to do so, we need insights both from the natural sciences, whose primary concern is the nature of risks and the methods used to reduce harm caused by disasters, and from the social sciences, which focus on methods of communicating risk. Naturally, then, our task is a multidisciplinary one. While rooted in social psychology, this volume brings together experts with backgrounds in sociology, civil engineering, information science, meteorology, and other fields, each of whom has contributed a report from the cutting edge of disaster risk communication research and practice. A brief overview of each chapter follows.

Chapter "Pitfall of Disaster Information" considers the pitfalls of disaster risk communication from the perspective of the "double bind," a well-known concept in the field of communication research. More specifically, this chapter suggests that this perspective can be used to gain a comprehensive, unified understanding of practical problems related to disaster risk communication, such as the general public's overreliance on disaster information and the division between experts and nonexperts. The specific initiatives described in subsequent chapters may be viewed as attempts to resolve the pitfalls created by the "double bind."

Chapter "New Approaches Toward Tsunami Risk Preparedness in Japan" takes up practical problems related to communicating the risks of a future Nankai Trough earthquake and tsunami, mentioned briefly above, and proposes a completely new and unique type of evacuation drill called a "single-person drill" as a solution. This approach fundamentally revises the way that evacuation drills have been practiced in Japan up to now—that is, by large groups of people at schools or in neighborhoods—and shifts the focus to individuals. The goal is to communicate disaster risk through evacuation drills in which participating community members feel personally responsible for and invested in what is happening.

Chapter "Using Computer Simulation for Effective Tsunami Risk Communication" describes the use of agent-based simulation to address the same issues as Chap. "New Approaches Toward Tsunami Risk Preparedness in Japan". A precise evacuation simulation based on the results of an exhaustive survey of community members is presented here, but the point of this simulation is not so much to produce highly accurate results as to use agent-based simulation—which has in the past been used by scholars of earthquakes, tsunamis, and evacuation behavior simply as an analytical tool—as a tool for communication between experts and nonexperts (community members) on the subject of disaster risk.

Chapter "Gaming Approach to Disaster Risk Communication: Development and Application of the "Crossroad Game"" describes efforts to communicate disaster risk by using gaming techniques. The "Crossroads" gaming technique discussed here is a tool that uses actual decision-making scenarios faced by disaster victims and responders during the Great Hanshin Earthquake of 1995 in Kobe to encourage discussion and consensus building among players. This chapter reports on the use of this technique in the area affected by the 2011 Great East Japan Earthquake, as a tool for survivors to communicate their disaster experiences to the next generation.

Weather disasters such as hurricanes and river floods are the subject of Chap. "Collaborative Community Weather Information". In developed regions such as Europe, North America, and Japan, information about the risk of these disasters is typically communicated to the public by specialized agencies, such as the Japan Meteorological Agency. However, some observers have pointed out the limits these agencies face in providing information about extremely localized or unexpected disasters. In response, effective initiatives have been launched to reduce damage through the collaborative production of so-called "local weather information" by community members and experts.

Chapter "Disaster Education Based on Legitimate Peripheral Participation Theory" discusses two projects aimed at changing the nature of earthquake response education in the context of debates in psychology, which forms the basis for this type of education. In the first project, observation stations (seismographs) comprising part of an earthquake monitoring network developed by experts were placed in elementary schools and managed by elementary students. In the second, members of the public worked to turn an earthquake monitoring station into an earthquake science museum. Both projects approach learning about earthquakes not as a process of communicating knowledge and techniques, but rather from the perspective of legitimate peripheral participation (LPP), that is, as participation in actual earthquake science research activities.

Chapters “Do Developed Countries Learn DRR from Developing Countries?” and “Imminent Warning Communication: Earthquake Early Warning and Short-Term Forecasting in Japan and the US” address examples from outside of Japan. Chapter “Do Developed Countries Learn DRR from Developing Countries?” discusses disaster aid in developing countries from the perspective of disaster preparedness education, taking up the 2015 Gorhka Earthquake in Nepal as an example. Disaster preparedness education is liable to be understood as the communication of knowledge and techniques from those who have knowledge to those who do not. However, developing nations do not always receive support in a unidirectional manner; to the contrary, this chapter points out that there is much to learn from developing countries, and raises questions about one-sided disaster aid.

Chapter “Imminent Warning Communication: Earthquake Early Warning and Short-Term Forecasting in Japan and the US” focuses on short-term earthquake forecasts, which have not been effectively utilized in Japan thus far due to concerns over liability for causing public panic, unnecessary preparations, or oversights occurring in relation to the release of such information. However, revelations that the Great East Japan Earthquake in 2011 and the Kumamoto earthquake in 2016 were directly preceded by smaller earthquakes that experts believe could have been used to predict the subsequent large earthquakes have drawn increased attention to the potential for short-term earthquake forecasts to reduce damage. This chapter uses examples from the United States and a comparison between the situation there and in Japan to discuss these potential and related issues.

Researchers and practitioners interested in reducing the damage caused by disasters have in the past devoted their energy to reducing negative outcomes caused by ignorance of disaster risks. This is, of course, important work, and will continue to be crucial in the future. However, we have already entered the stage where it is essential to recognize that knowledge or excess knowledge of disaster risk can also lead to negative outcomes, and we need to advance research and on-the-ground initiatives to deter these outcomes. It is my hope that the unique disaster risk communication methods presented in this volume will contribute to this goal.

Katsuya Yamori

Contents

Pitfall of Disaster Information

From the Perspective of Double-Bind Theory

Katsuya Yamori

Abstract Even though disaster information provided on its own completely isolated from all other communication can be logically comprehended, in reality it is meaningless. In other words, all disaster information exists in the form of disaster risk communication, messages from one person to another person. When disaster information is viewed as communication (messages), it faces the numerous major difficulties associated with messages in general. "Double bind" is the name given to one of these difficulties by Gregory Bateson, a cultural anthropologist. A double bind arises in a situation in which the recipient of a message, and also in fact the sender of the message, faces a dilemma due to inconsistencies and conflicts arising between a message and its meta-message. Many of the problems involving disaster information in modern societies are thought to closely resemble the double bind. This chapter proposes that these problems can be conceptualized as three major conditions: double bind that reproduces "waiting for information," double bind that reproduces dependence on government and experts, and double bind that reproduces an objective and unequivocal view of disaster information. A few innovative challenges to overcome difficulties of double bind, under the central concept of "joining and sharing" information, are introduced.

Keywords Disaster information · Risk communication · Double bind · Message · Meta-message

1 Meta-Messages

Even though disaster information provided on its own completely isolated from all other communication can be logically comprehended, in reality it is meaningless. In other words, all disaster information exists in the form of disaster risk communication—messages from one person to another person. When disaster information is viewed as communication (messages), it faces the many major difficulties associated

K. Yamori (✉)
Kyoto University, Kyoto, Japan
e-mail: momo-san@mx5.canvas.ne.jp

K. Yamori (ed.), *Disaster Risk Communication*, Integrated Disaster Risk Management,
https://doi.org/10.1007/978-981-13-2318-8_1

with messages in general. "Double bind" is the name given to one of these difficulties by Bateson (2000), a cultural anthropologist who developed an independent theory of psychiatric communication. Put simply, a double bind arises in a situation when the recipient of a message—and also in fact the sender of the message, as described later—faces a dilemma due to inconsistencies and conflicts arising between a message and its meta-message.

First, we need an explanation of meta-messages. We will give a slightly modified and simplified introduction using the example on which Bateson himself focused. Assume that we have two puppy dogs. Sometimes they bite or scratch each other with their claws, and we sometimes cannot distinguish this behavior from a genuine attack. However, since they normally do not inflict fatal wounds on each other, we can recognize that they are actually just playing.

There are two parallel levels of messages in this play behavior. The first message is "This is an attack. I'm going to bite you!" If this message did not exist, the play (pretend fight) itself would be meaningless. However, at the same time, there is another parallel meta-message, namely, "This is just play, and is not an actual fight." The reason why the second message is a meta-message is, first, that it is not explicit: it is an unspoken message that is not at the center of the communication. Another reason is that it serves the function of giving meaning to the first message from a different level than the first message itself.

The fact that messages are constructed of the two components of message and meta-message is the same in communication between humans, where it is normally even more apparent. For example, people in the Kansai region, Japan, including Osaka and Kyoto, often say "you really are an idiot" to people who have not been careful and have gotten hurt, or to people who they think "have gone too far." In this case, there is, first, the explicit message of the literal meaning of the words. This is the critical message that "people who fail are fools (idiots)" and "it is not good to get too worked up about things." However, at the same time, there often also exists a parallel meta-message that expresses the nuances of concern and affection ("try to be more careful next time") and further advances a sense of respect along the lines of "taking things that far certainly takes some courage."

An extremely important point here is the inconsistency and conflict that arises between the message and the meta-message. In both of the above two examples, if we look at the logical relationship between the message and the meta-message when placed side by side, then we would have to say that they are mutually inconsistent and conflicting. This is because we have two mutually inconsistent messages occurring at the same time, with "this is a fight/this is not a fight" in the former example, and "you are a fool/you are not a fool" in the latter example.

However, the inconsistency and conflict is normally overcome somehow due to the two messages being separated into an explicit level (the level of the "figure" or the literal level) and a meta-level (the level of the "background" or the context level). Moreover, they are able to produce extremely subtle communication through the mutual assumption that there is another message. In other words, the play is fun because it is partly serious, and "idiot" is employed as a word that is humanizing because it is recognized as partly critical and partly affectionate.

However, there may also be cases where this inconsistency and conflict comes to the foreground. In the former example, the play may unexpectedly develop into a true fight. In the latter example, the person who was called an "idiot" may regress into thinking "I really am an idiot." When this happens, the person who called the other an "idiot" is put into an awkward position, which often results in a tit-for-tat exchange with replies such as "don't give me that attitude." The meta-message is lost, and the primary message takes with its literal meaning.

2 Double Bind

The inconsistency and conflict between the message and meta-message may develop into the problem where one dominates (normally the message), as seen above. However, a more profound case noted by Bateson involves the exposure of a true inconsistency held by both the message and the meta-message while each still maintains its efficacy, and the recipient of the message is unable to proceed in any direction. This is the essence of a double bind. In such cases, the inconsistency remains exposed as an inconsistency instead of communicating deeper meaning and complex nuance since following either the message or the meta-message separately sends into two different levels. This is called a double bind because a person who is bound by the two mutually contradictory acts of following the message (the first bind) and following the meta-message (the second bind) becomes unable to act. In other words, a double bind is regarded as the state where communication itself becomes dysfunctional due to the conflicting relationship between message and meta-message.

An example that is often raised as a typical double-bind state in double-bind theory is the communication seen in relationships between an overprotective parent and an overdependent child. For example, there are university students who are barely able to attend a lecture if their parents do not organize it for them, such as by saying "if you do not take this subject in first year, it will be a problem later" (we have heard that such people actually exist). This is the kind of parent who would call the university and make inquiries because they are concerned whether or not their child is attending lectures properly (the authors have actual experience with this). In essence, this is a relationship where the parent cannot be independent from the child and the child cannot be independent from the parent, which frustrates both parties, but they are unable to overcome the situation, and so it continues interminably in this form.

In this kind of relationship, the parent attempts to instruct and command the child with phrases such as "you are a university student and you should be more independent" and "resolve it by yourself" (in fact, this kind of communication is also widespread). Consider what a child who receives this message should do. The first message, in other words the explicit message, is the literal instruction or command to become separate and independent from the parent. However, the important thing is that this message is simultaneously accompanied by the meta-message "obey my direction or command to become independent" (in other words, "do not become independent"). This kind of meta-message is produced by the facial expression, way of speaking, body language, or other actions of the parent (such as becoming worried and contacting the university to check on the class attendance of the child while giving such instructions at the same time) that are expressed together with the message of "become independent." As can be easily understood, at this time, the child obeying the message means disobeying the meta-message, and obeying the meta-message means disobeying the message. Thus, the child faces a dilemma.

It is important that this double bind, which paralyzes the child, also binds the parent who issued the message. If the child takes the attitude of relying on the parent as usual, they will disobey the first message. However, even if the child begins acting freely and independently, this means disobeying the meta-message and the parent will become concerned, thinking "is my child really ok?" While the double bind of becoming independent from parents/separating from children is a double bind between the message and meta-message in this sense, in terms of the fact that both the message sender and message recipient are bound, this can also be called a double double bind. Because of this, both parties actually mutually support each other, even while they are attempting to separate from each other, and the state of mutual dependence is protracted.

3 Double Binds Involving Disaster Information

Although the above explanation of double-bind theory was somewhat lengthy, it was necessary in order to clarify the basic problem that the authors have become aware of as a problem of disaster information faced by modern societies, including Western and Japanese societies. Stating our conclusions first, many of the problems involving disaster information in modern societies, particularly the problems noted by Katada et al. (for example, Katada 2006; Katada et al. 2005) using the terms "over-dependence on disaster experts," "heavy reliance on government-centered disaster management style," and "waiting for information resulting in delay of evacuation" are thought to closely resemble the double bind which we found in the mutually dependent parents/children. The main actors in communication where double bind occurs are the disaster experts, government agencies, and mass media (which correspond to the parent in the above example), who are the main sources issuing and disseminating disaster information, and the general public (which correspond to the child in the above example), who receives that information.

Before we begin looking at each one of the specific issues in the following sections, we will first take an overall look at the authors' idea of being able to conceptualize disaster information as a double bind. For this, it is essential to focus on the meta-messages that are produced in the communication on disaster information. For example, let us consider the information "XX river is at risk of breaking its banks due to heavy rain since last night. Immediately evacuate to your designated evacuation site." This message could be accompanied by a variety of meta-messages as described below, and the author thinks that this is in fact the case.

One is the meta-message of "do not evacuate until you receive this message. In other words, wait for a message." Needless to say, this is thought to be the main cause of the set of problems identified as "waiting for information." This point is described later in Sect. 4. Furthermore, the following meta-message also exists. The meta-message is "in this world, there are people like us who have the role of issuing (or disseminating) such messages, and there are people (everyone else), who have the role of receiving such messages and acting based on their content." This is the meta-message that recreates the sender/recipient relationship (government/citizens relationship), which combines overprotectiveness with overdependence. This point is described in Sect. 5.

Furthermore, the meta-message of "disaster information is something that objectively and unequivocally identifies the state of nature and the state of society" (that is, it should be information from which all possible ambiguity and inconsistency has been removed) is also produced. For this, the authors recently proposed a disaster risk communication technique that takes the form of a disaster game called "Crossroads" (see chapter "Gaming Approach to Disaster Risk Communication: Development and Application of the "Crossroad Game"" of this book, and Yamori 2007, 2010), which takes a form that reproduces the inconsistency and conflict between the various people involved. This point is described in Sect. 6.

4 Double Bind that Reproduces "Waiting for Information"

4.1 "Waiting for Information" Causing Slow Evacuation

Although the quantity of disaster information has become more abundant in Japan in recent years, it fosters an attitude of waiting for disaster information among the residents of a region, and is actually becoming an obstacle to appropriate and rapid evacuation. This problem of "waiting for information" was most clearly stated and advanced together with empirical data by Katada et al. (Katade et al. 2005; Katada laboratory 2003). For example, Katada et al. (2005) conducted a study of the tsunami evacuation in Kesennuma City, which was affected by an earthquake off the coast of Miyagi prefecture that occurred in May 2003. In this earthquake, it was reported that the actual evacuation rate was less than 2%, despite it being a situation where a tsunami was expected to arrive based on previous cases.

In this report, the questions that Katada et al. (2005) focused on in particular were the reason why residents did not evacuate and what they were doing during that time. Katada et al. (2005) emphasized the following points based on the fact that Kesennuma City, which was the target of the survey, is a tsunami-prone region, and that any tremors of seismic intensity 5-upper and above, in Japanese Meteorological Agency Scale ranging from 0 to 7, are expected to be accompanied by the arrival of tsunami. When respondents who had answered "I thought about evacuating but I did not evacuate" were asked for their reason for not evacuating, 54.6% (largest group) answered "because I received information that there was no tsunami damage." This information was the message that "although there has been a change in the sea level, there is no risk of tsunami damage," which was disseminated 12 min after the earthquake occurred.

Katada et al. (2005) noted that many of the residents "spent the 12 min until the 'no tsunami damage' information arrived continuously waiting for information on the tsunami, such as tsunami warnings, while preparing to evacuate" because approximately 80% of respondents were watching TV since the earthquake occurred during dinnertime at 6:24 pm. More specifically, "the decision to evacuate relied excessively on information" is viewed as a problem. Ushiyama and Imamura (2004) also noted that over 70% of survey respondents answered "waiting for tsunami warnings or evacuation guidance" in regards to the same incident. "Waiting for information" is also observed in other cases of disasters in Japan. For example, in cases such as the September 2003 earthquake off the coast of Tokachi (Yoshii et al. 2004) and the September 2004 earthquake off the southeast coast of the Kii peninsula (Kawata 2006; Kuroda 2008), it was noted that many of the local residents were in the state of "waiting for information" and "waiting for instructions" from the local government or other authorities, and this delayed the tsunami evacuation.

This phenomenon also occurred in the Great East Japan Earthquake. Of course, since the disaster-affected regions in the Great East Japan Earthquake were concentrated in the coastal areas of the Pacific shore of the Tohoku region, where the sense of caution regarding tsunami is thought to be higher relative to other regions in Japan, the sense of having to "wait for information" in this region is less pronounced compared to the previous examples. More specifically, it has been reported that more than 50% of people started evacuating (without waiting for information) immediately after the tremors ceased (for example, Working Group for Investigating Tsunami Evacuation Measures, Committee for Advancing and Investigating Disaster Management Measures, Central Disaster Management Council 2012a). However, looking at the entire country overall, the trend of delaying evacuation due to "waiting for information" was pronounced even for this major earthquake. For example, the Working Group for Investigating Tsunami Evacuation Measures, Committee for Advancing and Investigating Disaster Management Measures, Central Disaster Management Council (2012b) found the following from a questionnaire survey conducted on respondents who were aware of the danger of tsunami inundation (11,962 people) from among residents of the regions, where a warning of a large tsunami was issued.

First, the most common response to the question, "Did you do any of the following immediately after the earthquake?", where multiple responses were allowed, was "Acquire information about the earthquake" (approximately 60%), and the second most common response was "Acquire information about tsunami" (approximately 30%), which far exceeded any other responses. Furthermore, the two most common responses to the question, "What was the reason for evacuating to escape from the tsunami?", where multiple responses were allowed, were "Heard or saw a warning of a large tsunami" (approximately 35%) and "Heard or saw a tsunami warning" (approximately 30%), surpassing the third most common response of "I thought a tsunami would come based on the intensity of the tremor" (approximately 25%). The above results clearly demonstrate that delays in evacuation due to waiting for information and directions, in which evacuation begins only after acquiring or receiving information (regardless of the method or form) also occurred during the Great East Japan Earthquake in 2011 as in previous cases.

4.2 The Double Bind Associated with the Message "Do not Wait for Information"

It is important to ask how we should view the factors that cause this "waiting for information." Is it caused by the great volume and abundance of disaster information in our information society, or is it caused by the loss of our sense of protecting ourselves by ourselves? Surely these are obvious reasons for "waiting for information." However, in order to solve the problem, we first need to ask the question of why and based on what mechanisms the abundance of information induces the attitude of passively depending on information instead of the attitude of actively utilizing information, and why it is so difficult to break free of this situation.

In this sense, the double bind involving disaster information related to advice and directions on evacuation is considered to provide a satisfactory explanation of the root of the problem of "waiting for information." More specifically, in many cases, with each repetition, the disaster information of "you should evacuate" that is issued by disaster experts and disseminated by the government and mass media continues to deliver the following meta-message to the people at the same time as, and ironically more powerfully than, the literal meaning of the message. That is the meta-message of "you should evacuate after receiving disaster information" and "need not evacuate unless you receive disaster information."

Of course, experts in disaster information and related government bodies are also gradually noticing this inconsistency and paradox. This is the reason why it has recently become common to hear messages such as "evacuate quickly without waiting for information" and "decide by yourself without relying on information." However, one thing that tends to deepen the problem is that the message "evacuate quickly without waiting for information" itself is again a form of double bind. The reason is that accepting the message of "do not rely on information" means relying on information issued by experts and the government (namely, the information of "do not rely on information"). Needless to say, this is exactly the same as the pattern of becoming even more interdependent in the double bind we looked at earlier, which inhibits the process of becoming independent from parents/separating from children.

Thus, we can ask the question of how we can escape from this double bind. The key to this can be seen in studies on double binds introduced later in Sects. 5 and 6. In this section, we outline only the important point constituting the key to resolving this problem, namely, the importance of fully identifying a double bind as a double double bind. As described earlier, a double bind does not only bind the recipient of the information. It also binds the sender of the information (experts, government, and mass media). Because of this, this kind of problem cannot be solved using approaches such as changing the content of the information into a form which is easier to understand or attempting to increase the literacy of the recipients. In fact, instead of solving the problem, these kinds of approaches actually become the causes of the problem. This is because the kindness of the sender actually creates the dependence in the receiver (taking the approach of waiting), and the sender of the information is also spoiled by (depends on) the dependent attitude of the recipient. This is the same scenario in which parents who clean up after their child while grumbling "you really are a completely hopeless child" are at the same time unable to separate from their child.

5 Double Bind that Reproduces Dependence on Government and Experts

Attention also needs to be paid to the fact that at the same time as issuing the message, a lot of disaster information simultaneously issues the meta-message of "in this world, there are people, like us, who have the role of issuing (or disseminating) such messages, and there are people (everyone else) who have the role of receiving such messages and acting based on their content." This is a meta-message that plays a major role in reproducing the relationship of overprotection and overdependence between the citizens who depend excessively on the government and experts for their own safety, and government and experts who paternalistically attempt to excessively control the safety of the citizens.

Of course, the sense of danger in such circumstances has already been pointed out. Debate over reviewing "self-help, mutual help, and government help" in the field of practical disaster reduction in Japan is one indicator of this. For example, looking at the tsunami evacuation scenarios described above, the debate over reviewing self-help, mutual help, and government help typically takes the following form. That is, residents are asked to re-acknowledge that they themselves are the final authority regarding the decision to evacuate, and an appeal is made to residents to evacuate earlier through their own judgment (self-help) instead of simply following the conventional system of evacuating under the guidance of an autonomous body (government help). At the same time, in this case, the residents in a neighborhood are expected to mutually urge each other to evacuate, to assist with the evacuation of elderly people, disabled people, etc. within their local community, and to back this up through advance training and drills with the participation of the residents (mutual help). In summary, the trend is to once again put residents, who have come to expect nothing more than a passive role, despite the fact that they are the ultimate subject of tsunami evacuation, into the position of the main actor, and to cast off their dependence on government and experts.

We will now look at two examples where this point is thought to be relatively concrete and important. First, the research of Katada (2006), in which a disaster prevention project involving researchers, government bodies, and residents was initiated using a custom-integrated tsunami disaster simulator as the core tool in the tsunami-prone region of Owase City in Mie Prefecture, similar to Kesennuma City described above, is rather evocative. Katada (2006) conducted a survey of the actual situation relating to the reactions of residents during an earthquake off the southeast coast of the Kii peninsula (two earthquakes occurred on September 5, 2004, one in the evening and one late at night), which the residents of Owase City happened to experience right in the middle of a series of initiatives promoting disaster prevention.

The point we should focus on is that, like the example of Kesennuma City described above, although the overall evacuation rate was low, the Minatomachi District, which lies on the coast, had the highest evacuation rate, followed by the Nakaimachi District, which neighbors the Minatomachi District on the inland side. According to Katada (2006), this was because many of the residents of Minatomachi passed through Nakaimachi as they were moving to their evacuation sites, and residents of Nakaimachi who saw this also evacuated. The fact that hearing and seeing people evacuating promotes evacuation in this way has also been confirmed in other cases, such as the Typhoon #23 disaster in 2004 and the Great East Japan Earthquake (2011). The existence of this kind of mechanism is also backed up by experimental research related to group behavior (Sugiman and Misumi 1988).

Because of this, Katada (2006) recommended appointing "lead evacuators" within autonomous disaster prevention organizations in the local communities. The lead evacuators are people who always begin evacuating quickly after an earthquake occurs while calling out to neighbors. In other words, people whose own evacuation behavior serves the role of information for other people. Furthermore, the fruits of this trial are well known as the "Miracle of Kamaishi" (Katada 2012) in the Great East Japan Earthquake.

"Tsunami Tendenko" is another such case, which became well known particularly after the Great East Japan Earthquake. "Tsunami Tendenko" is an oral tradition of the Sanriku Coast region of Tohoku, which frequently experiences extensive tsunami damage (including the Great East Japan Earthquake), and refers to everyone agreeing in advance to evacuate "separately and individually" when faced with the threat of tsunami occurring in order to prevent entire families from being lost together (Yamori 2014). As discussed in Yamori (2014), the meaning and function of this tradition is extremely multilayered, and it would be naive to understand this simply in terms of this primary meaning. However, in our context, the tradition is clearly an example that plays the role of breaking free from the attitude of waiting for information and the problem of dependence on government. This is because it states that individuals should evacuate autonomously and rapidly upon feeling a large tremor, without waiting for or depending on information, etc. to be issued by the government or by mass media.

In summary, the points that should be studied based on these cases is the insight that as long as the construct where a person acts based on the trigger of disaster information is maintained, it is not possible to escape from the construct of recipients and senders of disaster information due to the side effect of the meta-message. An alternative solution that has been proposed to replace this construct is the idea of (the behavior of) regular people themselves serving as disaster information. As described above, “lead evacuators” are people who themselves serve as information. “Lead evacuators” are nothing more or less than an attempt to escape from the negative consequences brought about by the construct of the double bind by taking on the active role of people mutually and cooperatively issuing disaster information to each other through their behavior (the act of evacuating).

6 Double Bind that Reproduces an Objective and Unequivocal View of Disaster Information

6.1 Is It Possible to Remove Ambiguity from Disaster Information?

There is also the following example of a meta-message, i.e., an implicit message generated by disaster information and attached to the message itself. This is the meta-message of “disaster information is something that objectively and unequivocally identifies the state of nature and the state of society.” Put another way, this is the meta-message that ambiguity, vagueness, conflict, and inconsistency should be eliminated completely in disaster information, and that disaster information should take the form of a single specification of the situation and directions for behavior, ideally, in an “if…then…” form, such as “the amount of precipitation per hour has exceeded × millimeters, and so a level 3 heavy rainfall warning is to be issued” or “evacuate if the dangerous water level is exceeded.” A system that groups together many pieces of information in this kind of format is nothing more or less than a disaster management plan and response manual. Of course, the vagueness and ambiguity contained in disaster information sometimes results in time being wasted checking information or undertaking inappropriate disaster response actions. Taken further, the fact that this also forms a hotbed for irresponsible rumors and gossip is common knowledge in the field of social psychology. The author has no objections regarding the recognition of these issues itself.

However, it is not the case that all uses of disaster information end with this view of information. In fact, with the growing tendency for disaster information to become removed from information related to the natural phenomenon itself and to become information related to the human and societal response (social phenomenon), it becomes more difficult to utilize it for an objective and unequivocal view of disaster information. For example, this becomes clear only if we attempt to compare the set of problems related to the creation and dissemination of emergency earthquake reports with the set of problems related to the issuing and dissemination of information related to the human and societal response to the emergency earthquake reports (see chapter “Imminent Warning Communication: Earthquake Early Warning and Short-Term Forecasting in Japan and the US”). It is clear that compared to the former, in the latter case many more problems of vagueness and ambiguity of information will unavoidably emerge due to differences between people (for example, differences in age and gender between people receiving the information) and differences between environments (for example, a regular home, school, hospital, public transport infrastructure, etc.).

The best example of this is the set of problems related to tsunami evacuation in the Great East Japan Earthquake in 2011. The tsunami information and the evacuation for this large earthquake created an abundance of extremely complex instructions (information and lessons), which are hard to be formulated in an unequivocal and decisive format, such as “A must be done/B must not be done.” Rather, we faced in the catastrophe many dilemmatic choices in the form of whether or not question, such as whether or not it is OK for the fire department to prioritize their own evacuation, whether or not it is OK to evacuate by car, whether or not it is OK to select a person in need whom to assist during the evacuation, whether or not it is OK to evacuate to the designated evacuation site, whether or not it is OK to move from one evacuation site to another, and whether or not so-called vertical evacuation is OK. For all of these, an abundance of mutually contradictory survey results and episodes has already been reported, when we attempt to specify a single correct answer (Yamori 2014). In summary, a wide variety of lessons, experiences, and mutually inconsistent ambiguous facts should be respected, formulated, and disseminated as useful disaster information, rather than unambiguous facts represented in an unequivocal manner.

6.2 Mutual Dependence of Disaster Information and Disaster Management

One more reason for looking at the third type of double bind is the mutual dependence of disaster information and disaster management. The idea of what disaster information should be (the view of disaster information) and the idea of what practices should be needed for disaster management (the view of disaster management) are mutually defined and reinforced. As suggested by the theory of situated learning (Suchman 1987), a wide variety of community practices in modern society, covering everything and not just limited to disaster prevention, are complex cooperative practices that involve a large number of people and mediating tools. The tools in a broad sense (with information also regarded as a type of tool) for easily visualizing this and making it easier for people to cooperate and coordinate, come into play at this point. Thus, the practices we speak of here (disaster management practices) and the tools that support them (including disaster information) as noted earlier mutually support each other in a positive sense, but form a mutually vexing chicken-and-egg relationship in a negative sense.

The important point suggested by this understanding is that there is not only a cause-and-effect relationship that is easily understood through common sense, in which we first have practices and then develop tools to suit those practices, but also the cause-and-effect relationship in reverse direction, in which the tools (disaster information in this context) shape the practices themselves (disaster management activities in this context). More specifically, there is a risk that while conventional disaster information focusing on objectivity and unequivocality generates the meta-message stating that other types of information are not disaster information and propagates it through the society, it might also unwittingly trivialize specific aspects of the disaster management practices.

Let us think more specifically and in more detail. For example, there is the information of images from weather satellites. Many tools are involved in this process, including the images themselves, the satellites that shoot the images, and the devices for sending and receiving the images. What counts here is that the use of these tools is inextricably linked with a widely shared assumption among people that there is an expertise related to developing and operating those tools, that creating weather information based on the images is an expert profession, that the mass media plays the role of widely distributing the weather information, and that general people are expected to wait passively for this information and to react to it in some way. The disaster information produced and disseminated in a series of experts' practices and the dichotomic view of disaster management practices, between active disaster experts and passive nonexperts, are two sides of the same coin. Therefore, utilizing this type of information creates a foundation that supports these kinds of disaster management while at the same time forming an obstacle to thinking about and implementing disaster management in a different way.

In addition, disaster management manuals or plans can also be viewed as another example of mutual reinforcement of disaster information types and disaster management styles. Information in the manuals and plans normally organize a large number of people into a system of fixed positions and roles, and allocating activities to people based on the powers and duties assigned to them within that system. This organization not only conveys an explicit message about concrete contents of positions and roles, but also an implicit meta-message of a managerial view of disaster reduction practices in which every practice should be prescribed and controlled properly without any ad hoc responses. Of course, from the perspective that this type of information, represented in manuals and plans, can be used to organize many human and physical resources (where otherwise they would be distributed randomly and disorderedly) and to visualize their stock and flow, this information greatly contributes to the disaster management practices in a format that differs again from the previously discussed weather satellite images. However, at the same time, this could also be said to close off other paths for organizing and executing disaster management in a different way.

As can be easily imagined, there is little meaning in questioning the relative merits of the two views of disaster management practices described above and the two types of views of disaster information that are adapted to them. Both are able to build separate worlds, and both have their respective strengths and weaknesses. The important thing is to discard the fixed view of disaster information (for the reason that it creates a fixed view of disaster management styles), to search for the existence of other ways, and to combine these with existing measures in order to create an abundance of options. In this sense, the "Crossroads" disaster education game developed by the authors (see chapter "Gaming Approach to Disaster Risk Communication: Development and Application of the "Crossroad Game"" and Yamori 2010) can be said to aim for a third path that is based on and yet different from the above two types of disaster information, which focus on objective and unequivocal description of natural or social conditions, respectively, and constructing disaster management based on these.

Put simply, "Crossroads" visualizes the inconsistency and conflict between the multiple mutually conflicting understandings, attitudes, and values, among various stakeholders, including both experts and nonexperts, related to natural and social aspects in disaster management. The game aims to produce meta-messages: multilateral gaming processes in which the players with diverse backgrounds engage (exchanging opinions, conflicting in some time, persuading and being persuaded each other, and creating and collapsing consensus) are in fact disaster management, rather than a unilateral flow of directions and resources. The fact that a lack of motivation to ask for universally correct solution in "Crossroads" (Yamori 2007, 2010) indicates that the disaster information that is circulated within the game is vague and ambiguous, unlike conventional, rather managerial style of disaster reduction practices. This is an unavoidable property, even if a negative score is obtained, as long as we assume existing views of disaster information, that is, existing top-down views of disaster practices.

However, as described earlier, nowadays there is demand for multi-streaming and diversification of disaster management practices, which equals multi-streaming and diversification of views of disaster information. As long as the specific and existing views of disaster information are adhered to, it is difficult to break free from the double bind that they create. We wish to point out the importance of continuously searching for the existence of new disaster prevention practices while constantly focusing on the double bind created by existing disaster information, while remembering that the disaster prevention practices in "Crossroads" are not exceptions.

7 Summary—Issues and Challenges

We finish this chapter by emphasizing the reason why we need to start thinking about the warning about double binds in disaster information now, and at the same time introduce the first proposals for specific directions for resolving these problems.

Sociology is said to be the "self-awareness" of modern society. In the modern world, which is described as a "risk society" (Beck 1992) and "disaster-prone age," the disaster reduction research that is expected to serve as the front line for studying and implementing measures against disaster risks is required not only to advance the core activities of acquiring and developing disaster reduction knowledge and technology, but also to take the perspective of recursively looking at the positions and functions that the research itself occupies (self-awareness) in a society that confronts natural disasters under the assumption of the fruits of this knowledge and technology. In other words, disaster reduction research demands not only acquiring knowledge and technology related to data on natural or human and social phenomena, but also having a clear awareness of the roles played by itself in the "risk society" of today.

For example, disaster reduction research produces information and technology (such as knowledge and measurement technology for seismic waveforms) related to earthquake risk evaluation (which is almost purely data-focused), as well as social information and technology that assumes this (such as the creation and dissemination of forecasting information and emergency systems for rapid earthquake reporting). However, this information and technology is accompanied by secondary risk. For example, there are cases where failure of the forecasts themselves, economic loss accompanying forecast information, or reaction of users to emergency earthquake reports may have occurred (see also chapter "Imminent Warning Communication: Earthquake Early Warning and Short-Term Forecasting in Japan and the US"). Looking at this from a different perspective, this means that "unease increases with our knowledge" (Murakami 1998, p. 21). Disaster reduction research therefore attempts to view the secondary risk created by our own activities as a secondary, or recursive, aspect of data, as well as to forecast and control this by incorporating information and technology related to it into our own set of standard techniques.

The above indicates that in modern societies, including, Western and Japanese society, the context that supports disaster information is becoming more complex, and the possibility of circulating disaster information using literal messages is decreasing. In other words, the importance of a message in terms of the consequences brought about by circulating that message as well as in terms of the human relationships that contribute to the circulation of that message is increasing. This indicates the need to focus on the meta-message as much as, or even more than, the message, and to pay attention to the double-bind condition that may arise from the conflict and inconsistency between the two. The need to study disaster information from this perspective is expected to grow even more in the future.

We have already a few innovative challenges to overcome difficulties of double bind. The central concept of such challenges is "joining and sharing" (participating in and sharing) disaster information. Borrowing the words of the aforementioned Murakami (1998), "joining and sharing" is a joint and collaborative initiative, participated by not only disaster experts but also nonexperts, that attempts to co-create shared disaster information that is truly effective and casts off the "double bind" through "knowledge by everyone," instead of increasing uncertainty and dependence through passively receiving information disseminated by experts and governments. This critical point is consistent throughout all challenges introduced in the following chapters, from chapters "New Approaches Toward Tsunami Risk Preparedness in Japan: Single-Person Drills with Elderly Residents" to "Imminent Warning Communication: Earthquake Early Warning and Short-Term Forecasting in Japan and the US", although each chapter adopts different types of tools, methodologies, and concepts to tackle with various problems, rooted in double-bind situation.

References

Bateson G (2000) Steps to an ecology of mind. University of Chicago Press
Beck U (1992) Risk society. Sage Publications
Katada T (2006) Research on the existence of social co-learning based on a disaster survey and the results. 2005 Report on the results of important research topics (Research Grants) of the Research and Studies Committee of the Japan Society of Civil Engineers. http://www.jsce.or.jp/committee/jyuten/files/H17j_04.pdf
Katada T (2012) Disaster prevention where people do not die. Shueisha
Katada Laboratory (2003) Report on survey of the evacuation of Kesenuma City residents during the Sanriku earthquake of 26th May 2003 (Rapid Report). http://dsel.ce.gunma-u.ac.jp/modules/newdb1/detail.php?id=8
Katada T, Kodama M, Kuwasawa N, Koshimura S (2005) Issues of residents consciousness and evacuation from the Tsumani—From questionnaire survey in Kesennuma City, Miyagi Pref. after the earthquake of Miyagiken-Oki, 2003, J Jpn Soc Civil Eng 789(II-71):93–104
Kuroda Y (2008) Tsunamis and problems directly facing cities and Towns. Yoshii H, Tanaka J (eds) Introduction to disaster crisis management, Kobundo, pp 50–54
Kawata Y (2006) Surviving giant urban disasters. Shinchosha
Murakami Y (1998) The study of safety, Seidosha
Suchman L (1987) Plans and situated actions: the problem of human-machine communication. Cambridge University Press
Sugiman T, Misumi J (1988) Development of a new evacuation method for emergencies: control of collective behavior by emergent small groups. J Appl Psychol 73:3–10
Ushiyama M, Imamura F (2004) Disaster information and residents at the earthquake in Iwate and Miyagi prefecture on 26 May 2003, Tsunami Engineering Technical Report, No. 21, pp 57–82. http://disaster-i.net/notes/200305e-qr.pdf
Working Group for Investigating Tsunami Evacuation Measures, Committee for Advancing and Investigating Disaster Management Measures, Central Disaster Management Council (2012a) Collected reference material of the working group for investigating tsunami evacuation measures, Central Disaster Management Council
Working Group for Investigating Tsunami Evacuation Measures, Committee for Advancing and Investigating Disaster Management Measures, Central Disaster Management Council (2012b) Results of a web survey of earthquake and Tsunami evacuation during the great East Japan Earthquake (Rapid Report), Central Disaster Management Council
Yamori K (2007) Disaster risk sense in Japan and gaming approach to risk communication. Int J Mass Emergencies Disasters 25:101–131
Yamori K (2010) Using games in community disaster prevention exercises. Group Decis Negot. https://doi.org/10.1007/s10726-011-9227-9Key:citeulike:8711838
Yamori K (2014) Revisiting the concept of Tsunami Tendenko: Tsunami evacuation behavior in the Great East Japan Earthquake. In: Disaster Prevention Research Institute, Kyoto University (eds.). Natural disaster science and mitigation engineering: DPRI Reports (Vol.1), Studies on the 2011 off the Pacific Coast of Tohoku Earthquake. Springer Verlag pp 49–63. https://doi.org/10.1007/978-4-431-54418-0_5
Yoshii H, Tanaka A, Nakamura I, Nakamori H, Mikami S (2004) Evacuation activities at the Tokachi-Oki earthquake, 2004. Report Commissions to Survey Earthquakes, Ministry of Education, Culture, Sports, Science and Technology

New Approaches Toward Tsunami Risk Preparedness in Japan

Single-Person Drills with Elderly Residents

Yingying Sun

Abstract The 2011 Great East Japan Earthquake and following predictions of a Nankai megathrust earthquake have aroused great concerns on tsunami risks among the Japanese people. Among those concerns, three types of negative attitudes toward tsunami risk preparedness were perceived: overly pessimistic, overly optimistic, and overly dependent. To alleviate these negative attitudes, this study proposes a community-based tsunami preparedness strategy, termed as "single-person drill" in local communities characterized with high-aging populations. During the drill, an elderly evacuee wears a global positioning system device and heads away from his/her home toward a designated evacuation shelter. A group of school students record the drill process. All of the drill records are then synthesized into multiscreen movies with the following contents: the evacuee's behavior, messages of risk communication between the evacuee and students, and GPS tracking information overlaid on a simulated tsunami animation. This study suggests that the single-person drill has the functions of applying small-scale artefacts (e.g., video cameras and GPS devices) to the utilization of large-scale artefacts (e.g., seawalls and evacuation shelters). It also functions to shift the focus of tsunami risk preparedness practice from the community level to the individual level. More importantly, the single-person drill is found to have changed negative attitudes regarding tsunami preparedness, transforming residents' self-views from those who would need help, to being someone who would take initiative in reducing tsunami risks.

Keywords Tsunami risk prediction · Elderly residents · Elementary school students · Multiscreen movie · Artefact · Practice · Identity

This chapter is a revised version of Sun (2014) and Sun et al. (2014b).

Y. Sun (✉)
Sichuan University, Chengdu, China
e-mail: sunying@scu.edu.cn

K. Yamori (ed.), *Disaster Risk Communication*, Integrated Disaster Risk Management,
https://doi.org/10.1007/978-981-13-2318-8_2

1 Introduction

The increasing intensity and frequency of natural disaster events, combined with the growing proportions of elderly adults, have presented new challenges for the global community. Specifically, in Japan, the latest major, destructive natural disaster was the Great East Japan Earthquake. Measuring 9.0 on the Richter scale and with a maximum seismic intensity of 7, this was the largest earthquake to hit Japan in recorded history (Cabinet Office of Japan 2012b). It originated in the Pacific Ocean, east of Miyagi Prefecture, at 14:46 JST on March 11, 2011. The earthquake generated a huge tsunami, with wave height exceeding 10 m in some areas, and a maximum run-up height exceeding 40 m (Japan Meteorological Agency 2011). According to reports by the Central Disaster Prevention Council (2011), the National Police Agency of Japan (2014), and the Reconstruction Agency (2014), current records list about 20,000 fatalities, and 263,392 people living in refuge facilities at the end of 2014. Among those fatalities, over 90% were killed by the tsunami, 64% were above the age of 60 years, and 46% were over 70 years old (Ushiyama and Yokomaku 2012). In addition, the total area inundated by the tsunami exceeded 560 km^2, which is comparable to the area of Central Tokyo (621 km^2). A total of 127,305 houses were destroyed, with 272,941 partially destroyed (Cabinet Office of Japan 2012b).

Regarding the financial prediction of disaster recovery, the Reconstruction Agency (2011) estimated that there were about 16.9 trillion yen in direct financial damage, and that over 23 trillion yen would be needed for reconstruction over the next decade. In terms of the current reconstruction situation, according to the report of Miyagi Prefecture, coastal facilities had achieved a 50% completion rate and channel facilities a 20% completion rate, with household reconstruction finally becoming feasible as construction land was elevated to a certain height to mitigate tsunami risks (Miyagi Prefecture 2019). The tsunami caused both Japan and society worldwide to reflect on the tsunami risk reduction measures that should be taken to protect lives and property from future disasters. This is an urgent issue in Japan's eastern coastal regions, where the likelihood of a catastrophic Nankai megathrust earthquake occurring within the next 30 years is estimated as 70% (Headquarters for Earthquake Research Promotion 2013).

Prompted by the Great East Japan Earthquake, the Cabinet Office of Japan immediately organized several new working groups specifically to discuss appropriate countermeasures for a Nankai megathrust earthquake. The reports presented by these working groups indicate that the worst-case scenarios of inundated areas would total 1,015 km^2, 1.8 times larger than those affected in the Great East Japan Earthquake. They also predict that 124 municipalities would be inundated by a 5-m tsunami, and an additional 23 municipalities would be inundated by a 10-m tsunami. The highest projected tsunami could reach 34 m, with up to 323,000 fatalities being possible (Cabinet Office of Japan 2012a). These reports have aroused a great deal of concern among the public at large.

This chapter discusses the negative effects of the latest predictions of a Nankai megathrust earthquake and tsunami on community residents. With regard to related countermeasures, the chapter is mainly dedicated to describing a community-based tsunami preparedness strategy, termed as "single-person drill," which aims to promote effective tsunami evacuations in local communities. This chapter reports on the specific content of the "single-person drill" by presenting its implementation process, the contributions made by each participant, and the production process of multiscreen movies as research outputs. Based on this content, this chapter analyzes the essential meaning of continuous tsunami risk preparedness, the formation and transformation of communities of practice produced by activities of tsunami preparedness, and the identities of its participants.

The remainder of this chapter is structured as follows. Section 2 discusses the characteristics of tsunami predictions and subsequent negative attitudes toward future tsunami preparedness among local residents. Drawing on previous studies on tsunami risk preparedness and the present social situation in local communities, the author explains why personal approaches toward tsunami preparedness are urgently needed. Then, Sects. 3 and 4 provide detailed information about the implementation process and results of the "single-person drill." Section 5 discusses theoretical considerations from the perspective of the production of artefact, practice, and identity, which are core concepts in the theory of community of practice. Section 6 analyzes the essential meanings of community-based and community-initiated action research in tsunami risk preparedness and risk reduction.

2 Risk Predictions

2.1 Characteristics of Risk Predictions

In disaster studies, scholars point out that there are two facets of risk prediction: the objective matter of damage due to natural disasters and the variable matter of how human society is affected by such damage (Sun et al. 2014b; Yamori et al. 2005). Regarding the former, predictions are usually stated in a universal expression encompassing natural attributes, such as "it is supposed that an earthquake with an intensity of 7 on Japan Meteorological Agency's scale will occur, and a tsunami with waves exceeding 30 m will arrive." Such statements have no influence on the occurrence of a natural disaster, as the latter have their own characteristics of motion, independent of human activities (Sun et al. 2014b). For example, the movement of tectonic plates or the sea will not change due to the release of predictions as to their future movement trajectory. In this sense, such predictions describe the "objective danger" of a natural phenomenon or its results on human society. Thus, this type of prediction can be seen as characterizing "a neutral danger" (Yamori et al. 2005).

Regarding the extent to which human society is affected by disaster damage, such predictions are usually presented as issues directly related to society, such as "the most serious estimations predict as many as 323,000 deaths." Such statements do influence the actual results of damage, because damage varies according to societal and human beings' preparation. If more people have negative attitudes toward tsunami risk preparedness after hearing such predictions, the damage could increase to exceed even the direst degree of predictions. To some extent, such predictions depend on humans' decision-making and behavior, and thus contain the characteristics of "an active risk" (Yamori et al. 2005).

Examples of efforts toward changing the second type of prediction have been seen in several municipalities in Japan following the release of the latest prediction of a Nankai megathrust earthquake. These efforts include the development of new countermeasures against tsunami risks, and initiatives taken in motivating community residents to take part in tsunami risk preparedness (e.g., furniture fixation, emergency kit preparation, and evacuation drill performance). Specifically in Kochi Prefecture, government agencies even published a third prediction type, which describes the potentialities to change the second prediction type. For instance, the contents of the third prediction state that "according to the current situation of our region, if the rate of earthquake-proof houses is 74%, and the rate of immediate tsunami evacuation is 20%, meaning that 20% of residents evacuate immediately after an earthquake, then there will be almost 42,000 deaths." Continuously, the third prediction suggests that "if the above-mentioned rates were both improved to 100%, then deaths due to building collapse and fire would be reduced to one-tenth of the worst-case prediction, and tsunami deaths would be reduced to one-sixth" (Kochi Prefecture 2013). In sum, to encourage more people to continuously participate in tsunami risk preparedness, an urgent issue is that of helping them recognize the two facets of risk predictions. Specifically, it can be argued that helping people to understand their capability to change the second facet of risk prediction—that is, the active risk (Yamori et al. 2005)—will help local residents regain confidence in tsunami risk preparedness.

2.2 Three Types of Negative Attitudes

Three types of negative attitudes toward tsunami risk preparedness could be perceived among community residents following the release of the latest tsunami predictions of a Nankai megathrust earthquake. The first can be described as "overly pessimistic," and can be seen in local reports containing much negative commentary pertaining to tsunami risk preparedness. An example of the latter would be "they say a giant tsunami will arrive here. If it really comes, can't we just sit on top of the highest nearby mountain and cast our fishing lines into the tsunami? I will not make any further effort to prepare for the tsunami according to their (tsunami experts') predictions." Elderly people have also made comments such as "when the tsunami really comes, I would like to be swept away with my house" (Sun 2014). Nakamura et al. (2013) provide statistical evidence in their report. They conducted a survey in Shimoda and Kuroshio, both located in Kochi Prefecture, where the highest tsunami of over 34 m is expected to arrive. Given that the entire population of both areas would be affected by a tsunami, Nakamura and colleagues randomly selected 120 adults as informants. The survey results showed that 10.8% of respondents (6.7% in Shimoda and 15.0% in

Kuroshio) answered: "because predictions are so serious, I've started to believe it is meaningless to further prepare for tsunami risk reduction." Such pessimistic attitudes toward future tsunami risk preparedness and risk reduction could be a major reason behind even more deaths than those forecasted by the latest, direct predictions of a Nankai megathrust earthquake and tsunami occurring.

The second negative attitude regarding tsunami risk preparedness can be described as "overly optimistic." When the current author interviewed residents about their opinions of serious predictions, some people said things like "I do not think a tsunami will come during my lifetime," "it does not make sense to think so seriously about the tsunami. What will be, will be," and "I think that such predictions of earthquakes and tsunamis are unnecessarily disturbing news to our general community residents." Similar opinions have also been seen in newspapers, such as "I have been told that tsunamis would come ever since I was a little child. Whatever happens, happens" (Mainichi Shimbun 2012). Again, such optimistic attitudes to tsunami risk reduction could be a huge deterrent to enhancing residents' coping abilities in a tsunami situation (e.g., by making them hesitant to prepare emergency kits, reluctant to participate in evacuation drills, and opposing the building of higher shelters).

The third negative attitude toward tsunami risk reduction can be described as "overly dependent" on other people's efforts. The following are representative comments of this kind of attitude that were collected during the current author's activities in community-based tsunami risk reduction: "There's nothing can I do about the predictions. I surrender. I hope that disaster experts and government agencies will give us a hand." Such comments indicate a strong dependency on disaster experts and government agencies. Further examples highlight this: "I hope that safe evacuation shelters located at higher places will be constructed as soon as possible," "I hope that routes to evacuation shelters and bridges on the routes will be upgraded to withstand earthquakes as soon as possible," and "I hope that evacuation routes will be widened." This kind of dependent attitude revealed an emphasis on asking other people to take responsibility for one's own life, which could, in turn, hinder the cultivation of coping abilities with tsunamis.

These three types of negative attitudes are interrelated. From the perspective of cognitive psychology, Yamori (2013a) argues that "when people are faced with titanic peril, they either become desperate or inclined to belittle the peril." This indicates that overly pessimistic and overly optimistic attitudes are two sides of the same coin. Furthermore, as overly pessimistic attitudes guide people to give up on taking action to tsunami risk preparedness, and overly optimistic attitudes make people belittle taking action, overly dependent attitudes are thus easily formed. The eventual result is that of relying on experts and government agencies. However, previous disastrous events have demonstrated that it is the local people who can actually protect and rescue themselves in emergency situations. For instance, about 80% buried survivors were rescued by local residents in the 1995 Kobe Earthquake (Kawata 1997). In addition, the limitation of official agencies' abilities to protect people was presented by the fact that 256 firefighters lost their lives when they were helping people to evacuate in the Great East Japan Earthquake (Sun et al. 2013b). In sum, those three negative attitudes toward tsunami risk preparedness could arguably cause more serious damage than the direst prediction, if appropriate measures cannot be taken.

Needless to say, disaster predictions must be calculated based on professional knowledge and thus, in general, can only be produced by disaster experts and government agencies. However, there are two types of messages in such disaster predictions (Yamori 2013b). One is the literal message pertaining to the prediction of a disaster, while the other is a meta-message implied by the prediction (Yamori 2013b). For example, a literal message says "a tsunami has been observed. Please evacuate to higher places immediately." The implied meta-message is that "people like us produce and provide the predictions, and people like you receive those predictions and follow them to take actions." Thus, if there are overly adversarial or overly dependent relationships between community residents and disaster experts or government agencies, disaster predictions will be presented as a neutral danger. In this way, local residents would be largely disempowered in their intention to take any countermeasures against tsunami risks.

From the above results of the literature review, it becomes evident that the problem of communication as a one-way transfer that emerged during the Kobe Earthquake still existed at the time of the latest predictions of a Nankai megathrust earthquake (Sun et al. 2014b). To cope with this problem, it is necessary to establish a method of "communication for mutual consensus-making" (Yamori 2013b). Specifically, disaster experts should transform their position from knowing risk and calculating neutral danger, to serving as partners who collaborate with community residents by sharing active risk. Community residents, in turn, would benefit from transforming their attitudes toward disaster predictions, from only considering predictions as an unchangeable neutral danger to proactively taking part in reducing active risk (Yamori et al. 2005). Thus, the current chapter proposes that communication for mutual consensus-making between researchers and co-researchers (e.g., local residents) be posited as action research.

Action research is a research method whose objective is to explore theory in relation to practice (Eden and Ackermann 2018). Lewin (1946) initiated and developed the method of action research, stating its emphasis on knowledge produced in the context of application. Yamori (2010) proposed that two standards have to be met prior to undertaking action research. One is that the given social situation has been significantly changed during the process of trying to improve it. The other is that both researchers and research subjects need to share the same concern for achieving a better social situation. The following sections present the specific contents of the action research applied to this particular study based on ethnographic studies.

3 Single-Person Drills

3.1 Ethnographic Studies

Over the past 6 years, the current author has performed ethnographic studies of tsunami risk preparedness in Okitsu Village of Shimanto Town, Kochi Prefecture in Japan. Okitsu is made up of three small communities: Omuro, Urabun, and Gobun. The only transportation line is Road No. 52, which is a classic example of a narrow, winding mountain road, and has more than 140 curves along its 18 km length (Fig. 1). The road can easily be shut down by a landslide or earthquake. Since the release of the Nankai megathrust earthquake prediction, and in order to ensure the transportation of people and goods between Okitsu and the outside world in an emergency, Road No. 52 has been widened and maintained accordingly.

A major challenge in Okitsu is that of the serious impact of being hit by a tsunami. Ever since 2005, Okitsu has started to take countermeasures such as building evacuation shelters, maintaining evacuation routes, organizing community kitchen events, and investing efforts into school disaster education (Sun et al. 2013). Nonetheless, community members were largely discouraged by the latest tsunami prediction report. According to the reports by Kochi Prefecture (2012), an earthquake of an upper-6 intensity on the Japanese Meteorological Agency scale will cause a tsunami that will flood residential areas within 15 min. The maximum run-up height of tsunami waves is expected to exceed 25 m. As mentioned in Sect. 2.2, these tsunami predictions have resulted in three negative attitudes (overly pessimistic, overly optimistic, and overly dependent) among community residents. To cope with these problems, the current author collaborated with Okitsu Elementary School to develop the single-person drill, an approach aiming to encourage people to take part in tsunami evacuation drills. The author served as a coordinator to some extent, for

Fig. 1 Okitsu village. *Note* The village is fronted by the Pacific Ocean. On both sides of the mountain, there are two fishing ports. The indistinct line that can be seen is the mountain Road No. 52. (Modified from Sun et al. (2014a), Copyright 2014 Journal of Disaster Information Studies)

example, by motivating people to participate in the drill, visiting community residents' homes, taking part in village events, and studying with students in disaster education classes.

3.2 Design of the Single-Person Drill

The process of conducting the single-person drill was designed by the current author and colleagues at the Disaster Prevention Research Institute of Kyoto University, in early 2012. The single-person drill has become a regular activity in regard to tsunami risk reduction in Okitsu village since then. During a single-person drill, a single evacuee, typically an elderly person with a global positioning system (GPS) device, heads away from their home toward a designated evacuation shelter. A group of school students record the drill process, as a lesson in disaster education. All of the drill records are then synthesized into multiscreen movies, which are collaboratively made by school students, tsunami experts, computer graphic (CG) experts, and researchers. These movies are used as disaster education tools at the Disaster Education Symposium held by Okitsu Elementary School. The participants include students, evacuees, community residents, government officers, researchers, and journalists.

Figure 2 shows the tools used in the single-person drill: two video cameras, one stopwatch, and one GPS device. During the drill, five students make up one group to support a single evacuee. Two students serve as video recorders, A and B. The other three students serve as time recorders, interviewers, and notetakers (Fig. 3). The students are called drill supporters. Fourteen fifth-grade and sixth-grade students at Okitsu Elementary School were divided into three drill supporter groups, named the Omuro group, the Urabun group, and the Gobun group. In 2012, the total number of elementary school students in Okitsu was no more than 50. Due to the progress of modernization over the past 60 years, increasing numbers of young people have left Okitsu for large cities. Gradually, this phenomenon has made Okitsu a typical village labeled as a high-aging community with fewer children. Hence, the single-person

Fig. 2 Tools used in single-person drills (Reprinted from Sun et al. (2014b) Copyright 2014 Journal of Integrated Disaster Risk Management)

Fig. 3 Performing single-person drills (Modified from Sun et al. (2014a), Copyright 2014 Journal of Disaster Information Studies)

drills were intentionally targeted at children and elderly people, both of whom are often recognized as vulnerable people in disaster risk reduction (Sun et al. 2014b).

3.3 Implementation Processes

There are four steps to conducting the single-person drill. The first is a brief introduction (Fig. 4), which confirms certain aspects, such as where the evacuee will take shelter during an earthquake, whether the evacuee should take an emergency kit, and which routes and shelters will be used. The three remaining drill steps are also explained. Following this, the evacuee receives a GPS device to record their evacuation route.

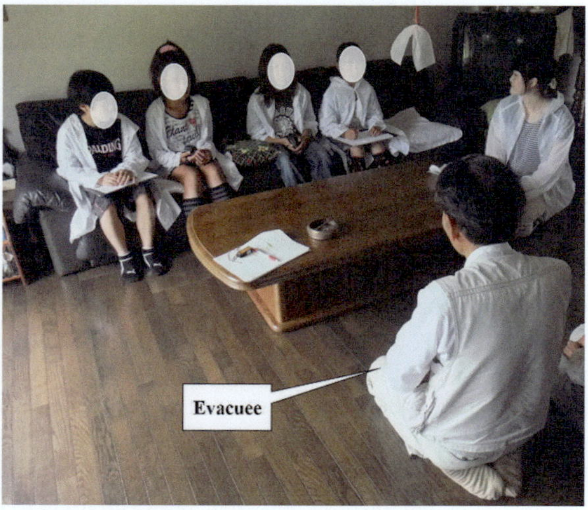

Fig. 4 Brief introduction and short interview during a single-person drill. *Note* The four students whose faces are masked performed as drill recorders. Photo taken by the author's colleague at Kyoto University

The second step involves a short interview right after the brief introduction (Fig. 4). The interviewer asks for information such as the evacuee's name, age, hobbies, and considerations for tsunami risk preparedness. This allows the evacuee and supporters to become somewhat familiar with each other before the actual evacuation.

The third step is that of the simulated evacuation. The time recorder says "earthquake" to announce the occurrence of an earthquake and the start of the evacuation. For the first 100 s, it is supposed that movement is impossible due to strong tremors. During this time, the evacuee takes shelter under a table or other stable furniture. After 100 s, the time recorder announces a second message: "the earthquake is over. Evacuate immediately," and the evacuee leaves. Some evacuees carry emergency kits with them containing items such as food, water, and flashlights.

During the evacuation, video recorder A (Fig. 3) stays close enough to the evacuee to capture the latter's facial expressions, while video recorder B maintains a certain distance in order to capture the surrounding situation. The time recorder uses a stopwatch to record the times required to reach significant locations, which are logged on a worksheet under "where" and "when" columns. For example, the time recorder might write "police station: 3 min 43 s," "intersection: 4 min 1 s," or "telegraph pole: 5 min 20 s." The interviewer walks alongside the evacuee, asking occasional questions such as "could you evacuate if this bridge were to collapse during an earthquake?"; "what do you think of the hand rails along this steep path?"; and "how are you feeling at this moment?"

The note taker also observes the surrounding situation and writes down information related to the drill. Example notes might be "the evacuee looks very tired now," "I think this is a dangerous place because the surrounding clay tiles may collapse," or "on a rainy day, it would be better to wear comfortable shoes that won't slip." When the evacuee arrives at an evacuation shelter, the time recorder announces, "arrival at destination," and all recording devices are turned off.

The fourth step is an interview survey. The evacuee and supporters retrace the route taken to check for dangerous places along the evacuation routes. When they return to the point of departure, the single-person drill is complete.

3.4 Products of Multiscreen Movies

The recorded results including the written notes, videos, and scientific GPS results of the single-person drills are synthesized into multiscreen movies by a collaboration of students, tsunami experts, CG experts, and researchers through three editing processes, as detailed below. Figure 5 shows the images from a created movie. The panels in the upper-left and bottom-right are camera videos taken from two angles. The upper-right panel shows two messages, given by the evacuee and students. One message displays problems found during the evacuation, and was given by the evacuee and students; the other message shows answers to these problems, and was given by students. The bottom-left panel shows GPS tracking information overlaid on a geographic information system (GIS) map that is part of a computer animation of a

Fig. 5 Multiscreen movies (Messages in the upper-right panel: "This road is too steep to walk." "Come on, you can do it.") (Modified from Sun et al. (2014a), Copyright 2014 Journal of Disaster Information Studies)

simulated tsunami. On the GIS map, a dot represents the evacuee's location during the drill, and a red line shows the evacuation route. A timer at the center of the movie presents the elapsed time during the drill. The contents of the four panels are displayed in synchronization with the timer. Detailed information of the editing processes is described below.

During the first editing step, tsunami experts collaborated with CG experts to create the upper-left hand panel content (close-up video recordings by camera A), the bottom-right (medium-distance recordings by camera B), and bottom-left (GPS data overlaid on a GIS map). Tsunami experts ran a tsunami simulation to calculate the movement of a tsunami in Okitsu based on the worst-case scenario published by the Cabinet Office. CG experts transformed the results of the tsunami simulation into a computer graphics animation. The animation was overlaid with the GPS data and superimposed on the GIS map in the bottom-left panel. Then, the timer was placed at the center of the movie. This allowed community residents easily to visualize the overall situation of the single-person drill. For example, this made it very easy to grasp how long it would take evacuees to arrive at a bridge, or where the first place to be inundated by a tsunami would be.

During the second editing step, CG experts and the author collaborated to place messages displaying problems found during the evacuation from the evacuee and students. The notes recorded by the notetaker and the interviewer were both used. The message format, suggested by the CG experts, stipulates that single messages are limited to two rows, with no more than 10 Japanese characters in one row. Original words and perceptions from the evacuee and students were rephrased into simple, clear command format. This allows for a timely presentation of message content and, furthermore, motivates viewer consciousness of the tsunami disaster and encourages students to offer more suggestions for tsunami risk reduction. Examples of edited messages are as follows.

The evacuee's original words were "I'm not afraid to go through tunnels on evacuation routes when I'm running to the evacuation shelter as fast as I can." In line with the multiscreen movie caption criteria, the researchers rephrased these words into "don't be afraid, just keep running." Similarly, another original statement of an evacuee, "this road seems too narrow for three people to pass at once. When an earthquake occurs, this might cause a bottleneck." This was rephrased into "walk in a single file on narrow roads."

During the third editing step, students, CG experts, and the author worked together to edit the messages from students and add them to the upper-right panel. The student supporters of the single-person drill greatly contributed to this editing. With reference to the unfinished movies, they carefully reviewed their supporting activities and responded to evacuees' messages by giving their opinions on the behavior and tsunami risk preparedness of evacuees' in Okitsu. For example, they provided encouraging messages to the evacuees and pointed out that it was necessary for the whole village to consider the problem of evacuating the elderly. Examples of their messages and responses are as follows:

Evacuee: Don't be afraid, just keep running.

Students: When evacuating at night, bring a flashlight.

Evacuee: Walk in a single file on narrow roads.

Students: Wider roads are safer.

Once all of the messages were added in, the multi-screen movie editing process was complete.

4 Results

To date, the students of Okitsu Elementary School have participated in two rounds of single-person drills, on June 26, 2012 and November 28, 2012, respectively. The drill comprised a total of 13 hours of integrated study periods. Specifically, 3 hours were used for tool operation training, 4 hours for drill implementation, 2 hours for editing messages, and 4 hours for presenting the multiscreen movies at the Disaster Education Symposium. Eight residents participated in the single-person drills, including two couples. Since the couples participated in the drill simultaneously, six pieces of multiscreen movies were produced. As the single-person drill approach specifically focused on vulnerable people (i.e., schoolchildren and elderly people), and illustrated and visualized the whole process of an evacuation, numerous mass media reporters were attracted to the activity scenes. Approximately, 35 journalists covered either the drill implementation or the movie presentations. Several schoolteachers were also interviewed about the single-person drill.

According to syllabus of Elementary Schools in Japan, the education time allocated to an integrated study period (e.g., class of disaster education, class of social practice) is limited. Therefore, students of Okitsu Elementary School cannot continuously take part in the single-person drills. The author thus stayed in Okitsu in May 2013, October 2013, and January 2014, in order independently to conduct single-person drills with community residents. Only one video camera was available for these drills, so the implementation process was somewhat simplified. To differentiate between the two types of drills, the original is referred to as the "basic single-person drill," and the simplified one as the "simplified single-person drill."

To date, 24 community residents have participated in the simplified drill, the results of which were recorded on paper. One side of the paper record presents all of the categories of information shown in the multiscreen movies, such as departure place, evacuation routes, evacuation shelters, time used for the drill, help among neighbors, and messages from evacuees and supporters. The other side presents two screenshots from a computer animation of a simulated tsunami overlaid with GPS data. These paper records were used to give feedback to evacuees.

After taking part in the single-person drill, some evacuees became clearly more interested in tsunami risk preparedness. For example, some participated in an annual evacuation drill conducted by Okitsu Volunteer Organizations for Disaster Risk Preparedness, or in evacuation drills for students at Okitsu Elementary School. During these drills, some participants went to different evacuation shelters to practice evacuating to secondary or tertiary shelters when primary shelters were inaccessible. Most participants recorded evacuation times using their own watches. For example, one of the single-person drill participants who had been "overly pessimistic" after hearing the latest prediction of a Nankai megathrust earthquake, actively participated in a community evacuation drill.

The time needed for each single-person drill varied with factors such as evacuee health, distance to an evacuation shelter, and weather conditions. Generally speaking, the full single-person drill from step one to four requires at least 30 min, and in some cases even required 4 hours. Evacuees with a high interest in participating in local events were willing to spend more time on step four. In contrast, others were unwilling to spend so much time on the drills due to work demands.

Students who served as supporters during the single-person drill gave presentations at the Disaster Education Symposium that convenes at Okitsu Elementary School, attended by all the students there, students from Okitsu Junior High School, community residents, government officers, journalists, and disaster experts. The following excerpt from one presentation is good evidence of the students' learning outcomes:

> We supported Tama in her single-person drill. Tama is 85 years old. She lives in the Urabun community. There are some dangerous elements near her house, such as clay tiles on roofs that might collapse after an earthquake. There are also dilapidated houses along her evacuation routes, so it is important to consider optional routes to the nearby evacuation shelter. It was raining heavily on the day of the evacuation, so moss on the steep evacuation routes got really slippery. It is necessary to undertake regular cleaning and maintenance of the routes. We also think elderly people should be sure to have comfortable, non-slip shoes, in case they need to evacuate.

This content was included in the multiscreen movies and, when read aloud by students, could be heard without reading the captions in the upper-right panel, which might be difficult for elderly individuals with poor eyesight. Thus, the presentation helped such viewers better to understand the movies, and also enhanced active risk communication between the generations.

In the presentation, evacuees also actively shared their experiences with other community residents. For example, Matsu, 65 years old, said "I was greatly encouraged by the students. I think I have shortened the time I require to get to the evacuation shelters by almost two minutes." M. Kajihara, aged 80, said "I forgot to turn off the gas in the kitchen before the evacuation." Such statements indicate participants' reflections on the single-person drill process and consideration of the appropriate measures to be taken during tsunami evacuations. Other residents, too, seemed highly motivated by the single-person drill. After the presentation, they immediately indicated their enthusiasm, saying they would like to participate again to help support the students. This highlights a relationship of mutual concern. After watching a tsunami computer animation, some residents indicated a better understanding of the damage a tsunami would cause, and wanted to know how long it would take a tsunami to arrive at their house. These changes that became apparent among local residents indicated that, given an appropriate approach such as the single-person drill, they would reconsider how their approach to disaster predictions. Accordingly, the objective of reducing "active risk" could be achieved.

5 Theoretical Considerations

In this section, some theoretical considerations are discussed regarding collaborative activities between researchers and co-researchers, by examining the significance of the single-person drills and the functions of multiscreen movies in tsunami risk reduction. During the single-person drills, three concepts emerged as important: artefact, practice, and identity.

5.1 Artefact

Artefacts can be physical objects such as computers or tables, or symbolic ones such as Halloween or Christmas. When these symbolic objects reappear frequently or repeat over time, they gradually become stable objects that acquire the same functions as physical ones (Sugiman 2006). In the field of disaster risk reduction, artefacts indicate infrastructure construction such as sea levees, evacuation shelters, highly earthquake-resistant buildings, and countermeasures against disasters such as hazard maps, community disaster maps, and evacuation wisdom. For example, organizations such as the Committee for Cooperation in Disaster Prevention, whose objective is to make policies to protect citizens' lives and properties, can be viewed as artefacts. Traditional wisdom regarding tsunami evacuations, such as the tsunami *tendenko* (*tendenko* refers to an "everyone for him/herself" mindset, which calls for a quick tsunami evacuation without waiting for others, not even one's parents or children (Yamori 2014)), can also be classified as an artefact.

The single-person drills in the current study flexibly applied small-scale artefacts such as video cameras and GPS devices to the utilization of existing, large-scale artefacts such as seawalls, evacuation shelters. Single-person drills also aided the production or replication of many new artefacts through the collaboration of various participants, such as the multiscreen movies and the Disaster Education Symposium. The single-person drills can thus shift comparatively large-scale artefacts from the whole community level to the personal, individual level in that, for example, when a resident actually climbs up to an evacuation shelter, the existential meaning of the shelter is changed in the resident's mind from that of a simple physical artefact to one of vital existence. Community residents who have never been to evacuation shelters may hold a mental image of a successful tsunami evacuation being impossible, and may have given up on the idea of evacuation. Actually visiting evacuation shelters with students could then engender the realization of the possibility of evacuation, thus transforming the mental image of large-scale artefacts. The drill also significantly contributes to the improvement of evacuation facilities, since the evacuation experience reveals the strength and weakness of such facilities.

5.2 *Practice*

Practice is the source of a community's coherence. Wenger (1998) proposes that in a community of practice, relationships between practice and community have three dimensions: the mutual engagement of participants, the negotiation of a joint enterprise, and the development of a shared repertoire. Wenger points out that the most crucial characteristic of communities of practice is a "shared practice," which indicates co-participation in practice or practical co-engagement. In the field of disaster prevention, a practice can be seen as efforts made by disaster experts to promote the resilience of local communities against disasters. Similarly, practitioners in local communities taking part in community kitchen activities so as to prepare for refugee life can also be seen as a practice. From the same point of view, when community residents collaborate with local government agencies to discuss the construction of new evacuation shelters or proposals to fix furniture inside their houses, this, too, can be seen as a community of practice.

Given that the single-person drill specifically emphasizes the importance of actual behavior by each participant, it also functions to shift the focus of tsunami risk reduction practice from the community level to the individual level. These drills can thus be considered as an effective way of overcoming the negative attitudes described above, by recasting tsunami evacuation from a general, nonurgent condition to a specific, concrete task that can be accomplished by individual residents.

The single-person drills promote the co-learning of tsunami risk reduction among intergenerational participants. The drill also combines researchers' active intervention in the community with neutral data collecting activities, such as that regarding evacuation behavior and facility effectiveness. Such collaborative practice between researchers and co-researchers is expected to overcome problems often seen in community-based disaster prevention activities, such as insufficient collaboration between community residents and disaster experts, and an over-dependency on infrastructure and experts.

The single-person drill involves people and artefacts that had not previously taken part in tsunami risk preparedness, whereby former strangers become partners in collaborative activities. As an example, producing the multiscreen movies combined evacuee behavior and academic activities to produce certain artefacts. The movies revealed information, such as frequently used evacuation routes and places, where brick walls should be removed to improve evacuations. The single-person drills thus produced or reproduced tsunami risk reduction practices and artefacts when community residents collaborated with government officers to upgrade dilapidated bridges or widen narrow evacuation routes.

Furthermore, from the perspective of taking initiative of practice, community residents actively measured their time from various current locations to the nearest evacuation shelter after the launch of single-person drills in Okitsu. As previously mentioned, the initial objective of the drills was to rebuild local residents' confidence in tsunami risk reduction, and to encourage people to participate in risk preparedness. Given the influence of the latest tsunami risk prediction and the impact of the simulated tsunami animation on the multiscreen movies, local residents gradually recognized the importance of "time" during an evacuation. Hence, the activity of measuring "time" could be seen as an effective practice generated by the cooperation between residents and researchers. Eventually, this practice could help to reduce the "active risk," one of the facets of risk predictions mentioned earlier.

5.3 Identity

Identity, in practice, is socially defined, not merely because it is reified in a social discourse of the self and of social categories, but also because it is produced as a lived experience of participation in specific communities (Wenger 1998). In the field of disaster prevention, when a Disaster Prevention Committee in a local community plans tsunami evacuation drills, various practitioners are expected to be involved. Examples include workers who construct evacuation shelters, community leaders who implement drills, and participating community residents. Through these practices, new problems or new evacuation methods are likely to emerge. For example, people who previously never participated in drills may join other family members and enjoy the drill very much. Thus, the identities of these practitioners are transformed because of the newly created relationships with artefacts and other members.

Disaster prevention practices should provide an appropriate stage for collaboration between community residents and disaster experts, which could eventually help to reconstruct and reproduce the identities of all participants (Yamori 2009). In the current study, the single-person drill was found gradually to have changed community attitudes regarding disaster, transforming residents' self-views from those who would need help, to being someone who would take initiative in reducing active risk. Disaster experts and government agencies also recast themselves, from people who knew the risks by calculating neutral danger, to one of the partners collaborating with community residents by sharing active risks.

Regarding identity transformations among students and community residents, it became clear that the single-person drills provided a good chance for students to participate in tsunami risk reduction by managing the drill process as disaster education. During the drills, students took detailed notes about the whole evacuation situation from their own perspectives. After the drill, they participated in producing encouraging messages to evacuees, and discussing the problems and shortcomings of the overall evacuation process. This kind of disaster education provides a new approach to fostering supporters, rather than mere survivors, in a disaster. Students who had never previously taken part in real disaster prevention activities, because these are usually conducted by adults, became important participants providing significant support roles, rather than token positions. The single-person drills provided them with leading roles that reproduced their identities in the context of tsunami risk preparedness. In addition, students performed the role of helping with other people's evacuation, which helped to change their stereotypical image of being "vulnerable" with regard to disaster risk reduction. In short, via the single-person drills, the students involved became true supporters as well as campaigners for community-based disaster risk reduction.

The single-person drills easily evoked awareness of such roles because the evacuees' every movement was recorded by video cameras. Moreover, when elderly evacuees were surrounded by young students, they realized that others cared for their safe evacuation, motivating them to recognize their responsibilities regarding tsunami risk preparedness. These are essential motifs in tsunami risk preparedness, but have previously tended to be neglected. In sum, the single-person drill can change elderly evacuees' mindset toward tsunami risk reduction and motivate them to take tsunami risk preparedness. For people who always had a high interest in tsunami risk preparedness, the verification of their behaviors using the sophisticated technologies found in disaster studies can arguably motivate them to think of more appropriate countermeasures against disasters and apply them to future activities.

For example, the abovementioned evacuee, Tama, who climbed up to an evacuation shelter, previously commented that she would rather be flooded away with her house because "it is really hard for me to evacuate." After the drill, with support from the students and researchers, her family members continued to encourage her, even complaining that "if you do not evacuate during a tsunami, we'll think that you don't love us." This is a typical means of disaster prevention based on a mutually concerned relationship. Six months later, Tama actively participated in the evacuation drills for students, climbing up to the evacuation shelter with her grandchild. According to video records taken by her family members, Tama reduced her evacuation time by almost 2 minutes. Furthermore, Tama declared that "from now on, I will go to the shelter every day just for a walk."

Weakness due to advanced age can reduce confidence in elderly people like Tama, causing them to give up on countermeasures against a future tsunami. However, the single-person drills specifically urged the elderly to play leading roles in tsunami risk reduction. This was found greatly to encourage the elderly by sending messages such as "do not surrender to a tsunami, and neither belittle its force," "do not rely on others, but think of countermeasures for yourself and your family," and "you cannot resolve all problems at once, so try to clear obstacles one at a time."

Regarding identity transformations, disaster experts and government officers calculate tsunami estimates by running simulations that rely on parameters such as the earthquake epicenter, its seismic intensity, the evacuation speed, and the reception rate of evacuation warnings. These studies focus on whole communities or multiple large regions, rather than individuals. However, as discussed in the previous section, there are limitations to applying these studies to promoting disaster prevention awareness, and to individual activities in particular.

Researchers in the single-person drills not only adopted the neutral position of distributing research results to community residents, but also became collaborative partners with community residents in considering appropriate and effective countermeasures. For example, evacuating alongside community residents frequently uncovered new facts that simulations do not account for, such as unexpected walking paces among the elderly, or behaviors related to helping others. Participation in the single-person drill and careful record-keeping thus provided the researchers with valuable data related to human behavior. In light of these findings, it can be argued that disaster experts and government agencies can similarly reproduce their identities by collaborating with community residents and transforming their positions from thinking only in academic terms to caring about the betterment of their co-researchers. As Yamori (2013b) states, the ultimate goal of action research in disaster risk reduction is to achieve a unique consensus among social members; indeed, the single-person drill in the present action research functioned to change the identities of both residents and researchers from a negative or neutral standpoint to active participation in tsunami risk reduction.

5.4 Functions of Multiscreen Movies

The multiscreen movies in this study were produced based on the collaboration of various participants, which provided a platform for linking and visualizing human behavior during tsunami evacuations and tsunami flooding. Owing to the collaborative efforts put into the production of the movies, two types of dialogues were achieved. One was that between disaster experts and community residents, and the other between the natural and human sciences. Here, dialogue refers to the dynamic transformational positions between neutral danger and active risk in promoting tsunami risk preparedness.

First, dialogues emerged between disaster experts and community residents. Tsunami experts and CG experts made professional contributions in editing the movies. Their responsibilities involved developing sophisticated tsunami simulation models, inventing tsunami animations with high accuracy, and editing the four synchronized parts of the movies. For their part, the evacuees presented the methods, routes, and evacuation times. All of these elements are vital parameters in tsunami evacuations. The video records and student notes were also flexibly used in the editing process. In short, all the materials used in the multiscreen movies were produced by the collaboration between disaster experts and community residents.

Ushiyama (2008) argues that local disaster prevention should not be restrained to just community-based activities, but must also be tested based on professional skills and developed to the level of scientific knowledge with a guaranteed practice. The single-person drills are examples of the latter. Although experts did strongly influence the editing process, the animation of predicted tsunami movement on the multiscreen movies would be dangerous or even damaging to the community residents without being backed up by accurate tsunami simulation calculations. The movies can thus be seen as artefacts that generated dialogue between actual and professional practices, as conducted by disaster experts and community residents.

Second, there emerged dialogues between the natural and human sciences. The contents shown in the multiscreen movies were produced by actual evacuee behaviors. For example, the upper-left and bottom-right panels displayed every movement during the evacuation. The message content shown in the upper-right panel was produced by evacuees and students during editing. The bottom-left shows evacuation methods and other elements that were the results of participants' decision-making. The movie contents were thus produced based on the human sciences.

In contrast, the design process of the movies is based on advanced technologies. Specifically, developing the models of tsunami simulation required applying the latest tsunami predictions provided by the Cabinet Office. The worst-case scenarios shown in the movies were scientifically calculated by setting coordinate systems, epicenter, mesh sizes, and various scenarios in advance. Such preconditions allowed for running tsunami simulations to predict damage over time. Sophisticated technologies in the field of disaster prevention and CG production were thus integrated with the movies. Although the importance of cross-disciplinary studies has been advocated for several decades, most research achievements have been limited to the pure disciplines of either natural science or social science. Across the studies that have adopted methodologies from various disciplines, such as statistics, computer science, and big data, human beings are usually treated as objective subjects instead of active participants.

Moreover, a problem with current disaster prevention methods is that, while many studies have been conducted from the perspective of the social sciences, their methodologies are borrowed from the natural sciences. Such studies clearly separate researchers and research subjects, so that the former can objectively observe the latter. However, if social science studies were to change their focus from natural to human phenomena, and consider only topics outside of the natural sciences, the social sciences could never keep pace with natural sciences (Yamori 2009). In this sense, the single-person drill approach employed in the current study functioned as a new action research methodology to combine knowledge from both natural and human science. The implementation of the single-person drill and the production of multiscreen movies can, therefore, be seen as an initial step toward tackling problems of how to apply cross-disciplinary methodologies from both natural and human science.

Further research on disaster risk preparedness from a cross-disciplinary perspective could further develop this new initiated methodology.

6 Conclusion

This chapter has discussed the collaborative activities of the single-person drill, based on the theory of communities of practice. It has also described the function of multi-screen movies from the perspectives of two research stances. It is here important to understand that the relationships between theory and practice in the human sciences are always mutually influenced. For example, through journalists' or researchers' descriptions and discussions of the single-person drills enacted in this research, the tsunami risk preparedness activities in Okitsu were significantly impacted by outsiders, from the perspective of Okitsu residents. Thus, theory (e.g., tsunami predictions) is not a fixed object, but, rather, changed by the development of practice. Consequently, it may be said that the functions of practice are to verify the rationality of theory and to reconstruct it.

Finally, it should be said that single-person drills cannot, of course, solve all the problems related to tsunami risk preparedness. While the drills emphasize collaboration between disaster experts and community residents, on some occasions the students and evacuees were put in subordinate positions. In addition, while focusing on personal behaviors and considering concrete countermeasures proved very effective in overcoming limitations in tsunami risk preparedness, difficulties still exist in the implementation processes. For example, one significant problem was that some elderly adults were incapable of evacuating independently, and it was very difficult to appoint specific people to aid them in this. In addition, even though the multiscreen movies were greatly influential in motivating people to participate in tsunami risk reduction, their production cost was relatively high, making it difficult to produce an evacuation movie for each evacuee. These problems led to modifying the single-person drill to a simplified form. More recently, with the newly constructed solar street lamps in Okitsu, plans for carrying out single-person drills during nighttime are under discussion. Therefore, at the present time, the current author's colleagues at Kyoto University are considering to modify the implementation process of the single-person drill to, first, adapt to the new environment in regard to tsunami evacuation. And second, to overcome limitations of the drills. With the continuously created artefact–practice–identity circle, the drills were expected to become one of the sound artefacts to tackle with tsunami risk in regions along the Nankai Trough.

References

Cabinet Office of Japan (2012a) Report on estimates of the distribution of seismic intensity and the height of tsunami waves of a Nankai trough earthquake (in Japanese), 1 June 2014. http://www.bousai.go.jp/jishin/nankai/model/

Cabinet Office of Japan (2012b) 2012 Disaster prevention white paper (in Japanese) 1 Octo 2018. http://www.bousai.go.jp/kaigirep/hakusho/index.html

Central Disaster Prevention Council (2011) Summary of the damage from Great East Japan Earthquake (in Japanese), 1 June 2014. http://www.bousai.go.jp/kaigirep/chousakai/tohokukyokun/1/pdf/3-2.pdf

Eden C, Ackermann F (2018) Theory into practice, practice to theory: action research in method development. Eur J Oper Res 271(3):1145–1155. https://doi.org/10.1016/j.ejor.2018.05.061

Headquarters for Earthquake Research Promotion (2013) Long-term evaluation of Nankai trough earthquake movement (second edition) (in Japanese), 1 June 2014. http://www.jishin.go.jp/main/chousa/13may_nankai/

Japan Meteorological Agency (2011) Seismic intensity in areas affected by the Great East Japan Earthquake (appendix No. 2) (in Japanese), 1 Mar 2014. http://www.data.jma.go.jp/svd/eqev/data/2011_03_11_tohoku/0311_shindo.pdf

Kawata Y (1997) Prediction of loss of human lives due to catastrophic earthquake disaster. J Jpn Soc Nat Disaster Sci 16:3–13

Kochi Prefecture (2012) The second report on estimates of the distribution of seismic intensity and the height of tsunami waves of a Nankai trough earthquake in Kochi prefecture (in Japanese), 20 Aug 2017 http://www.pref.kochi.lg.jp/soshiki/010201/nannkai-3.html

Kochi Prefecture (2013) Damage estimates of a Nankai megathrust earthquake in Kochi prefecture (in Japanese), June 2014. https://www.pref.kochi.lg.jp/soshiki/010201/higaisoutei-2013.html

Lewin K (1946) Action research and minority problems. J Soc Issues 2(4):34–46

Mainichi Shimbun (2012) Nankai megathrust earthquake: a great shock by the predicted highest tsunami waves over 20 m, analysts suggest "upgrading evacuation shelters to higher places" (in Japanese). Mainichi Shimbun

Miyagi Prefecture (2019) Progress of disaster reconstruction and recovery, 20 May 2019. http://www.pref.miyagi.jp/soshiki/dobokusom/doboku-sintyoku-press.html

Nakamura I, Amano A, Fujiyoshi Y, Mizukami T, Nakamori H, Amano N, … Tanihara K (2013) Residents' consciousness and response to damage predictions of a Nankai megathrust earthquake: survey results in Shimoda and Kuroshio by digital broadcasting workshop (in Japanese). In: Paper presented at the 15th conference of Japan society for disaster information studies, Tokyo

National Police Agency of Japan (2014) Tohoku Earthquake Tsunami damage and police measures (in Japanese), 1 June 2014. http://www.npa.go.jp/archive/keibi/biki/higaijokyo.pdf

Reconstruction Agency (2011) Fundamental reconstruction policy after the Tohoku earthquake Tsunami (Reconstruction Head Office's decision) (in Japanese), 1 June 2014. http://www.reconstruction.go.jp/topics/000056.html

Reconstruction Agency (2014) Number of domestic refugees due to the Great East Japan earthquake (in Japanese), 1 June 2014. http://www.reconstruction.go.jp/topics/main-cat2/sub-cat2-1/20140425_hinansha.pdf

Sugiman T (2006) Group dynamics (in Japanese). In: Sugiman T (ed) Group dynamics in communities. Kyoto University Press, Kyoto, pp 19–86

Sun Y (2014) Action research to promote tsunami risk reduction: ethnographic approaches to disaster education and tsunami evacuation. Kyoto University

Sun Y, Yamori K, Kondo S (2013a) Disaster education based on community of practice: a case study in Okitsu community. J Integr Disaster Risk Manag 3(1):92–106. https://doi.org/10.5595/idrim.2013.0056

Sun Y, Yamori K, Tanisawa R, Kondo S (2013b) Consciousness of disaster risk and tsunami evacuation: a questionnaire survey in Okitsu, Kochi prefecture. J Nat Disaster Sci 34(2):127–141

Sun Y, Kondo S, Miyamoto T, Yamori K (2014a) New approaches to tsunami reduction: single-person drill and multi-screen movie (in Japanese). J Disaster Inf Stud 12:76–87
Sun Y, Yamori K, Kondo S (2014b) Single-person drill for tsunami evacuation and disaster education. J Integr Disaster Risk Manag 4(1):30–47. https://doi.org/10.5595/idrim.2014.0080
Ushiyama M (2008) Local residents' participatory activities in the "field" of interdisciplinary research (in Japanese). In: Proceedings of the 45th integrated symposium of Japan society for natural disaster science, pp 51–52
Ushiyama M, Yokomaku S (2012) Features of disaster fatalities (in Japanese). J Disaster Inf Stud 10:7–13
Wenger E (1998) Communities of practice: learning, meaning, and identity. Cambridge university press
Yamori K (2009) Human science of disaster management (in Japanese). Tokyo University Press, Tokyo
Yamori K (2010) Action research: action-oriented human science (in Japanese). Shinyosha, Tokyo
Yamori K (2013a) Change estimates into a lie: how to support this "change" (in Japanese). J Inf Stud 11:14–19
Yamori K (2013b) Improving disaster risk communication: a paradigm shift in disaster information research (in Japanese). Minervashobo, Kyoto
Yamori K (2014) Revisiting the concept of tsunami *tendenko*: Tsunami evacuation behavior in the Great East Japan Earthquake. In: Kawase H (ed) Studies on the 2011 off the pacific Coast of Tohoku Earthquake. Springer Japan, Tokyo, pp 49–63
Yamori K, Kikkawa T, Aziro T (2005) Risk communication through games (in Japanese). Nakanishiya, Kyoto

Using Computer Simulation for Effective Tsunami Risk Communication

Michinori Hatayama and Fuko Nakai

Abstract With the lessons learned in the Great East Japan Earthquake in 2011, the Japanese Cabinet Office recalculated some existing disaster assumptions. As a result, the Nankai Trough has been assessed as being at risk for a huge earthquake and tsunami. Evacuation plans have been reconsidered in each Japanese administrative sector to be more practical for saving resident's lives. However, two problems arise concerning tsunami evacuation planning. First, such a disaster creates an infinitely complex situation because of the many factors involved. Second, it is difficult to verify whether any plans will actually work. Therefore, we cannot fully demonstrate the feasibility of these plans. To cope with these difficulties, computer simulation is well known as an effective way to verify plans. In this chapter, an effective process to propose feasible options on tsunami evacuation for residents by combining two approaches will be discussed. The first is a computer simulation that calculates a large variety of evacuation situations to find their most effective behaviors. The second proposes practical methods for testing the behaviors proposed by the results of simulation. The process appears to be effective in developing improved evacuation plans for residents.

Keywords Evacuation plan · Tsunami · Multi-agent simulation (MAS) · Disaster risk communication · Geographic information system (GIS) · Feasibility study

1 Introduction

One of the lessons learned in the Great East Japan Earthquake in 2011 was that it is extremely important to consider the largest possible mega earthquakes and tsunamis from every possible angle (Cabinet Office, Government of Japan 2011). Japanese Cabinet Office recalculated some existing disaster assumptions. As the result, the Nankai Trough, located in the Pacific Ocean off southern Japan has been assessed as

M. Hatayama (✉) · F. Nakai
Kyoto University, Kyoto, Japan
e-mail: hatayama@imdr.dpri.kyoto-u.ac.jp

Nagoya Institute of Technology, Nagoya, Japan

K. Yamori (ed.), *Disaster Risk Communication*, Integrated Disaster Risk Management,
https://doi.org/10.1007/978-981-13-2318-8_3

being at risk for a huge earthquake and tsunami. The most serious area, Kuroshio in Kochi Prefecture is threatened with maximum 34 m tsunami (Cabinet Office, Government of Japan 2013). After the assessment was announced, some media reported that many people were giving up evacuation plans, feeling that there is nothing they could do in the event of such a disaster. Giving up any possibility of evacuating is a serious problem because people's lives may be saved if they do not give up hope in preparing for a disaster.

Since the backgrounds, evacuation planning has been reconsidered in each Japanese administrative sector to be more practical for saving resident's lives. However, two problems arise concerning tsunami evacuation planning. First, such a disaster creates an infinitely complex situation because of the many factors involved. For example, the type of hazard, occurrence time, and the weather can generate numerous scenarios. Second, it is difficult to verify whether any plans will actually work. There are too many scenarios to practice, and certain phenomena, such as a tsunami and cracks in the ground, cannot be generated for real-world simulation. Furthermore, human psychological factors can make the situation worse because of interactions among residents that affect their evacuation environment, such as a traffic jam on the evacuation course. In summary, we cannot fully demonstrate the feasibility of the plans. To cope with these difficulties, computer simulation is well known as an effective way to verify plans.

In this chapter, an effective process to propose feasible options on tsunami evacuation for residents by combining two approaches is discussed. The first is a computer simulation that calculates a large variety of evacuation situations to find their most effective behaviors. The second proposes practical methods for testing the behaviors proposed by the results of simulation. The process appears to be effective to percolate them as improved evacuation plans to residents, too.

2 Process to Create Feasible Options

In order to create feasible options, disaster risk communication in which residents lived in tsunami prediction area take part is essential and not only the results but also the process reaching to them should be focused. The CAUSE model (Rowan et al. 2010) denoted the five fundamental communicative goals of risk communication as follows:

C: Confidence in partners

A: Awareness of hazard, solutions, structure of government

U: deep Understanding

S: Satisfaction with solutions

E: Enactment or change

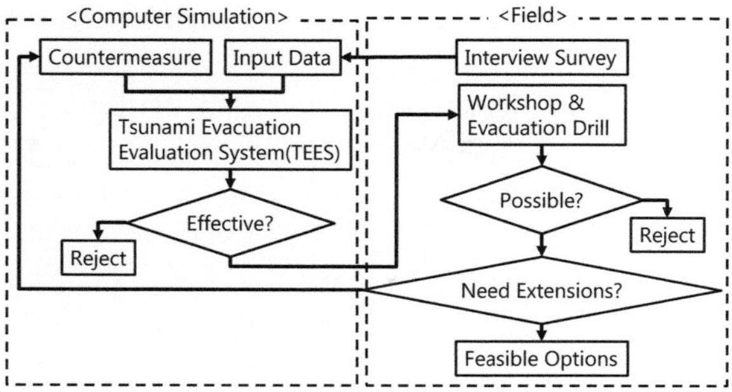

Fig. 1 Collaboration of computer simulation and fieldwork for disaster risk communication

However, it is hard for residents to aware, understand, and evaluate the current risks and the options to avoid the worst-case scenario. Given this factor, we assert to introduce computer simulation on tsunami evacuation as a tool to provide them the image of their current status and the effect of countermeasure. In this proposal, you have to pay attention that the solution is NOT generated only by computer simulation. Computer simulation generates only candidates of solutions, and residents will find their feasible options as their solution through a process of risk communication.

Figure 1 shows the concept for this research, displaying the process of deriving some feasible options by the collaboration of computer simulation and fieldwork (Nakai et al. 2015). Practical ideas must first be extracted from some of the residents, from the researchers, and from officials.

The first step is an interview investigation to find out what evacuation plans are in residents' minds and to determine which evacuation problems make the residents pessimistic.

The second step is computer simulation. The simulation model is based on residents' evacuation plans, and then we can determine existing evacuation problems by looking at collective evacuation behaviors. After the simulation, the agents' log data and tsunami evacuation simulation are visualized as an animation by GIS. Using this simulation process may give us insights into what is effective.

The third step is discussing the effective options that might serve as candidates for practice by means of workshops and evacuation drills.

Three results are generated from the field perspective. The first is to identify the impossible options, those completely rejected by the residents. The second is to identify those that are possible but that would need further work before they can be put into practice. The third is to practice some of the options and improve them through further practice.

3 Tsunami Evacuation Evaluation System (TEES)

Tsunami Evacuation Evaluation System (TEES) is developed to supply simulation results in both numeric values and animations that describe evacuation actions, and share evacuation action processes among participants to make tsunami evacuation plan in each local community with a high workability.

For evacuation actions, the research shall evaluate by comparing current understandings with when a certain measure has been taken. As mentioned above, not only comparison of the resulting numeric values will be carried out; rather, the process will be reviewed, and the cause will be examined to evaluate the actions. Though various possibilities are available as scenarios for measures, this research will extract the anxieties of local residents when they think of evacuation, and concerns felt through the process of awareness surveys in advance as "matters to deliberate." By focusing on this, this research shall carry out simulations, and analyze and examine this. Verifying upon putting matters extracted from real local societies into simulation enables to handle reality-oriented problems, and is thus a critical aspect of this research.

The overall flow of TEES is shown in Fig. 2. TEES is designed with two subsystems, together having five functions. The first subsystem is Geographic Information System (GIS), which has four functions: (1) setting the simulation field, (2) managing the results of the simulation in a database, (3) evaluating the evacuation behaviors according to tsunami prediction information, and (4) visualizing through an animation based on a temporal axis. The second subsystem is a Multi-Agent Simulator (MAS) with the fifth function: (5) calculating the dynamic evacuation situation.

3.1 Geographic Information System (GIS)

Base map data to set the simulation field and tsunami hazard data to evaluate the evacuation behaviors are needed in GIS.

Road data which means roadside line data and road network data are essential roadside line data which are used as input data for MAS as the possible environment where the evacuee agents can move and road network data are used to navigate the

Fig. 2 Tsunami evacuation evaluation system (TEES)

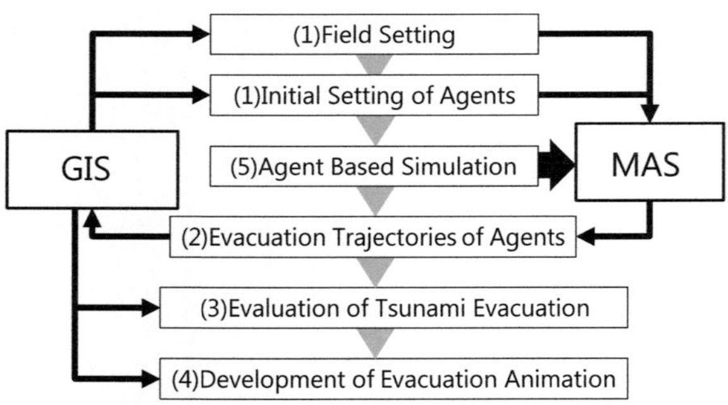

evacuee agent. The other base map components, such as buildings, coastal lines, rivers, and administration border, are used in making a visual for the background of the simulation.

Tsunami hazard data are used for the evaluation of the evacuee's behaviors according to tsunami prediction area and tsunami arrival time to the target area after earthquake.

3.2 Multi-agent Simulation (MAS)

With agent simulation being capable of observing behaviors of the overall system in a bottom-up manner, it is beneficial in discovering matters from various processes that occur during evacuation.

In this simulator, human agent is provided as a model for each person with the attributes that will determine their behavior as follows: original location, destination (safety spot), via-points, family connection information, transportation speed according to the age, starting time for evacuation, and car usage. The via-points are used if the agent wants to go to help the other agents, such as his/her family. If the human agent connected with the other agent by family connection information, they move as a group in which speed is slower than a normal person and faster than child or elder person. Starting time is expressed as an elapsed time after earthquake happens and included earthquake-shaking time.

4 Implementation to Case Study Area

4.1 Case Study Area

A case study area is Mangyo District in Kuroshio, Kochi, Japan, in which population is 600 with 250 households, which is located along the sea (Fig. 3). In the damage estimation report of the Nankai Trough Earthquake published in 2012, the biggest tsunami was assumed 14 m above sea level and reaching about 22 min after the earthquake. However, there are no hills around this district and about 15 min is needed to the closest hill in a non-disabled individual's walking speed. Therefore, a tsunami evacuation tower of 12 m above sea level had been built before Great East Japan Earthquake. Because it is not enough height to the evacuation spot for the 2012 assumption, a new evacuation tower whose height was 18 m above sea level was build next to the 12 m tower in 2014. The capacity of the new tower was 300 persons, which was not enough for all residents in Mangyo.

This case study is carried out after the publication of the 2012 assumption and before the construction of the new tower. As the first step, semi-structured interview survey for residents was carried out between the period of November 21, 2012 and

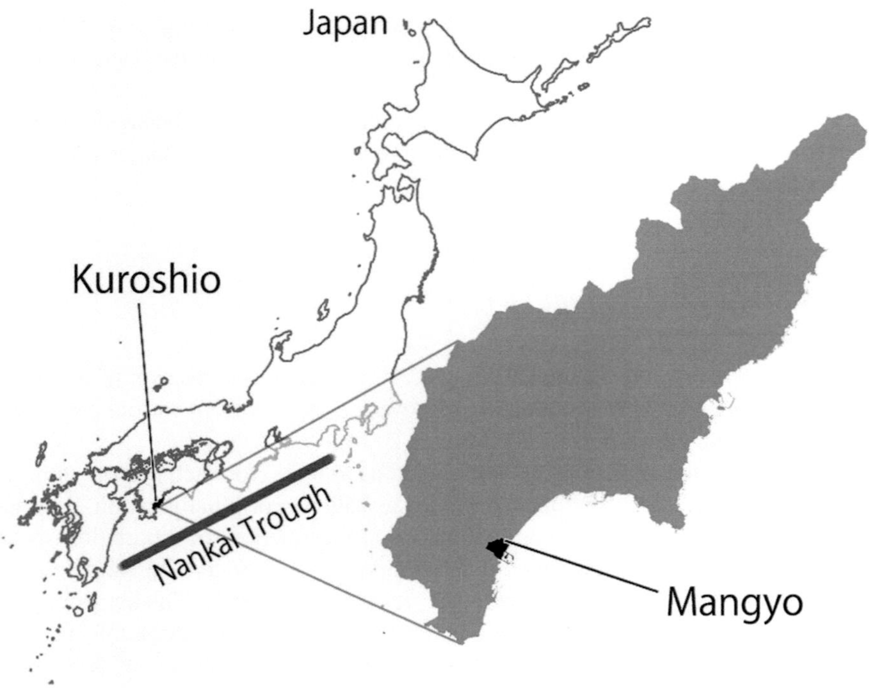

Fig. 3 Location of Mangyo, Kuroshio, Kochi prefecture, Japan

April 18, 2013. We finally obtained 296 responses including all households at that time in Mangyo.

4.2 Interview Survey

The interview items were designed to apprehend people's evacuation plan: original location (e.g., the location of their home or their office), planned destination location (i.e., the location of the shelter), the desired means (i.e., transportation, route), with whom they would evacuate, and when (i.e., the available time for preparing to evacuate).

The survey had two purposes. The first was to obtain data for determining the behavior model used in TEES, and the second was to discover the measures in people's minds by asking about their evacuation plans face to face.

Figure 4 shows the distribution of the shelters including the number of expected persons to evacuate. With the results from the interviews, the evacuation plans in the residents' minds were denoted. The residents claimed their concern of evacuation as follows:

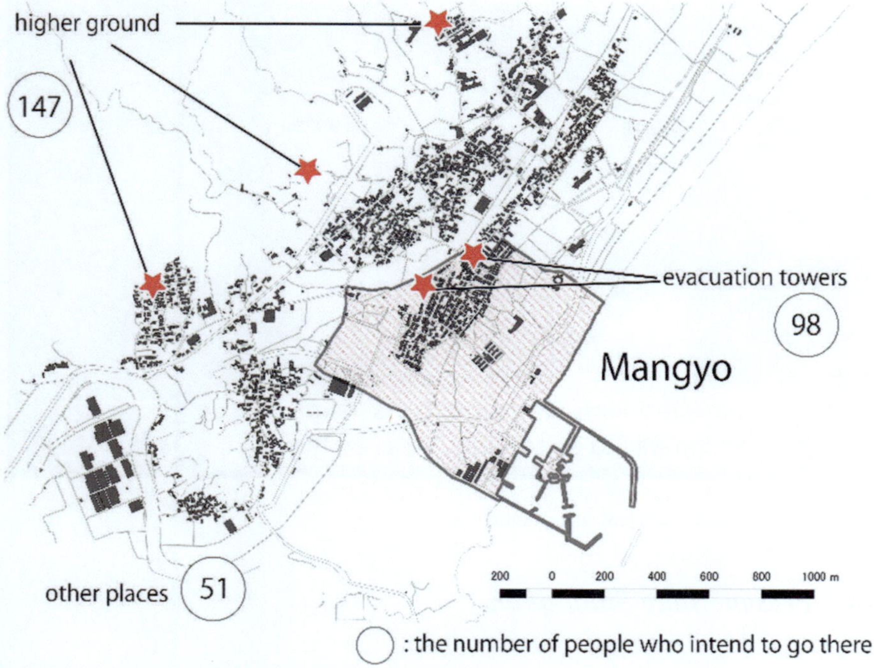

Fig. 4 Distribution of the shelters including the number of expected persons to evacuate. *Data source* Nakai et al. (2015)

(1) There is not enough time to prepare before evacuation.
(2) Most residents wish to go to hillsides where they can move to higher places according to the level of earthquake but they worried that the hillsides are located too far from where the residents live.
(3) Most residents wish to evacuate by car but they were worried about the delay due to traffic congestion.
(4) Normal people wish to support the disabled before tsunami but they were worried about the time limitations.

5 Proposal of Options Using TEES

To realize TEES, the base map provided by the Geospatial Information Authority of Japan and the tsunami hazard map based on one of the tsunami cases hypothesized by the Japanese Cabinet Office (Central Disaster Management Council, Cabinet Office, Government of Japan 2013) are implemented. This tsunami case is indicated as a most serious situation for the target area because the arrival time for tsunami waves is sooner than the other cases.

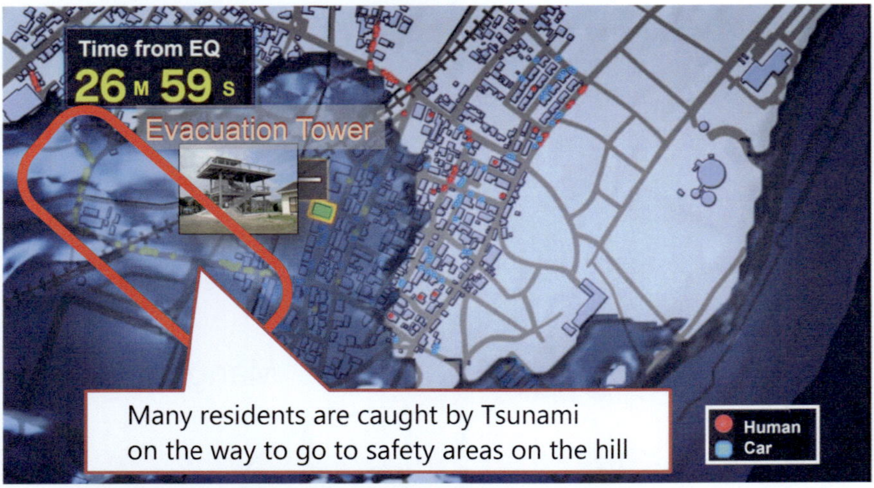

Fig. 5 A screenshot of current state simulation

5.1 *Present State Simulation*

As the second step, agent-based evacuation simulation according to the results of survey to the residents (hereinafter, refer to "present state simulation") was carried out and the issues of their concern were evaluated. The result showed that their concern could lead to the failures of evacuation. Especially, it pointed that many residents were caught by tsunami on the way to go to hillsides (Fig. 5).

5.2 *Proposed Options Based on TEES*

Based on the results of "present state simulation," three countermeasures were denoted as follows:

(A) Residents start their evacuation as soon as possible after experiencing the shaking due to earthquake without waiting for the tsunami early warning.
(B) Residents do not go to support disables who live on the opposite sides from their evacuation point even if disables are their family.
(C) Basically, residents don't use cars in evacuation. If they use cars, they share the cars according to their plan made by themselves before disaster.

However, TEES showed that only these countermeasures could not decrease the number of affected persons very much. An additional countermeasure was proposed based on the movement of the residents on the way to go to their planned evacuation points provided by TEES as follows:

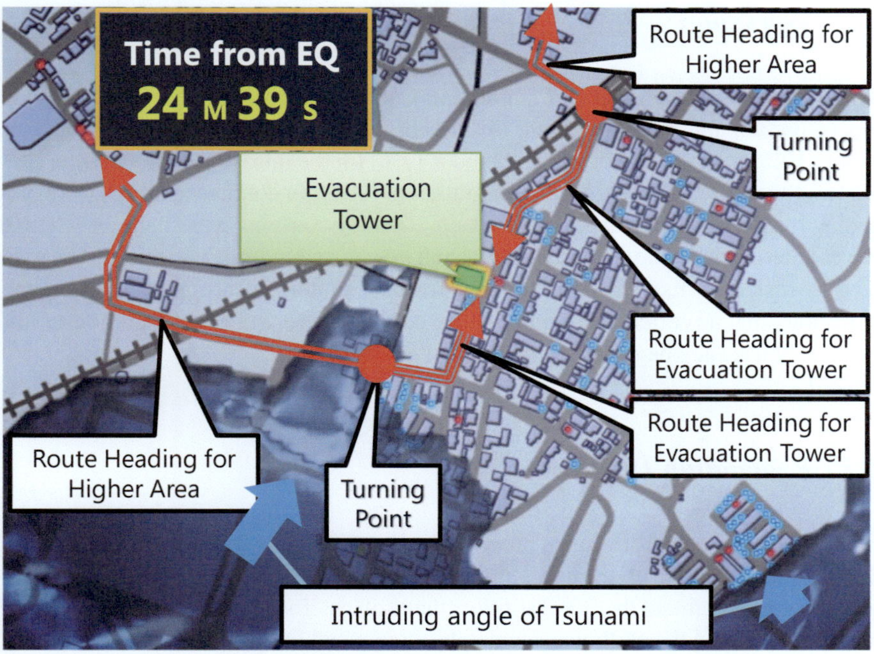

Fig. 6 Location of two turning points and evacuation routes

(D) Residents change their destinations to the evacuation tower on the way to intended evacuation place taking into account the passage of time from the earthquake.

To reach the hillsides from the district, residents have to pass either of two points which are shown as the turning points in Fig. 6. Residents who would pass the turning point 1 after 10 min or the point 2 after 13 min from the earthquake could not reach their evacuation places. However, if the countermeasure D would be introduced, the number of persons who would be caught by tsunami was decreased very much and if all of countermeasures would be introduced, almost all the evacuees could reach the evacuation places before tsunami. Before proposal of these countermeasures to residents, it was necessary to consider the capacity of the evacuation tower in the case. TEES showed if the decision time would be set at earlier than 10 min in the point 1 and 13 min in the point 2, it was possible that the evacuation tower would overflow its capacity because of the greater number of residents gathering there.

6 Field Activities to Verify the Feasibility of the Options

As the third step, workshop and evacuation drill were conducted as field activities by residents, local officials, and researchers.

6.1 Workshop

The workshop on the results of simulation has been offered three times. The number of participants was 80 (about 14% of the residents) for the first workshop, 60 (about 10%) for the second, and 100 (about 17%) for the third. Summary of interview survey, the results of "present status simulation," and four countermeasures were presented by the members of research group. Then discussions were held on whether the countermeasures could accept or not and how the countermeasures could be realized by residents. Especially, in order to realize countermeasure A, the participants confirmed that it is effective to improve the earthquake resistance of their houses and prepare emergency bags including survival goods.

6.2 Evacuation Drill

Evacuation drill according to the proposed countermeasures was implemented twice. The purpose of the drill was to help residents discover and experience the countermeasures and evaluate the feasibility for themselves. To test countermeasure D, the elapsed time from the earthquake had to be announced for the evacuees. Fortunately, in Japan, the time can be calculated from the combination of earthquake and tsunami early warning. In the first drill, one of the residents who first went up the evacuation tower shouted the elapsed time to the evacuees around the judgement place where they would turn back to the evacuation tower or not. In the second drill, the role of instruction was given to an electric bulletin board that showed the elapsed time. Based on the instruction, residents judged for themselves whether or not they should accept the countermeasure.

7 Discussion

The countermeasure D proposed by computer simulation was practiced in the evacuation drills as an actual experience. This process was important for the residents because it verifies the feasibility not only by computer simulation but also from the residents' perspective.

After the drill, the residents had many comments. Some of the residents (Group I) said, "I would follow this countermeasure if my family's lives would be saved." Some residents (Group II) are hoping for an advanced discussion of these countermeasures for realizing reliable instruction. However, one of the residents said, "I feel some resistance against turning from the intended shelter because it takes more time to turn and head to the evacuation tower"; another said, "I don't want to obey the instructions. I will evacuate based on what I think."(Group III) Summary is as follows:

- Group I have accepted countermeasure D as their feasible option.
- Group II have not accepted it yet but agreed one of the candidates of their option.
- Group III have not changed their mind and confirmed their original measure.

Except for this, we got some interesting reaction after the field activity as follows.

Some residents relocated their house to the place near their family. It was related to countermeasure B. Computer simulation showed that both individuals would be caught up by the tsunami in mutual help activities heading opposite from the direction of evacuation. Thus, visualizing the pair of rescuing and rescued individuals presented that the distance to move can be shortened by altering pairs. But the feasible option for the issues was not redevelopment of the pairs but relocating the supporters.

Some residents started walking to their evacuation spots as their daily work. It was related to countermeasure C. They understood the possibility to survive a tsunami and importance to know the current walking capabilities through communication.

These actions were the results to achieve five communication goals in CAUSE model of risk communication.

8 Conclusion

This research has proposed the collaboration of computer simulation and fieldwork and has revealed that it was an effective communication method in terms of considering a way to survive an extraordinarily catastrophic natural disaster. The most important idea was introducing the evaluation methods proposed by the residents themselves. Furthermore, the simulation, which is based on their actual intention, encouraged the residents to have more ownership as one of the outcomes.

However, a serious problem remains "I know what to do, but it is too difficult to do." The truth of this statement was revealed by the 2014 Iyonada Earthquake, which measured 6.2 on the Richter scale in the Seto Inland Sea of Japan at 02:06 JST on March 14, 2014. Mangyo also suffered from a big quake. According to the investigation that the authors implemented after the disaster, 73% of the residents in Mangyo evacuated by car (Sun et al. 2017). What actually happened here was completely different from the result of the interview survey. We may interpret their actions as reflecting their real intention because during the interviews, participants mumbled that they wanted to evacuate by car if they could. This was in response to the general opinion that they should not use their cars for evacuation, which was favored after the Great East Japan Earthquake. Given these real-world results, how to account for such a problem in a new plan requires a new search for a solution (Nakai et al. 2017).

References

Cabinet Office, Government of Japan (2011) Report of the committee for technical investigation on countermeasures for earthquakes and Tsunamis based on the lessons learned from the 2011 off the Pacific Coast of Tohoku Earthquake. Web Page of Disaster Prevention Information, Cabinet Office, Government of Japan. http://www.bousai.go.jp/kaigirep/chousakai/tohokukyokun/pdf/Report.pdf. Accessed 15 Oct 2018

Central Disaster Management Council, Cabinet Office, Government of Japan (2013) Second Report of Meeting for Mega Quake Model in the Nankai Trough, Tsunami Fault Model Version (Nankai torafu no kyodai jishin moderu kentoukai, dainiji houkoku, tsunami dansou moderu hen). Cabinet Office, Government of Japan. http://www.bousai.go.jp/jishin/nankai/taisaku/pdf/20120829_2nd_report01.pdf. Accessed 15 Oct 2018 (in Japanese)

Nakai F, Hatayama M, Yamori K (2015) Integrating computer simulation and the fieldwork to discover possible alternatives for Tsunami evacuation. In: Proceedings of the 48th Hawaii international conference on system sciences, Hawaii, USA, pp 200–207

Nakai F, Mochizuki J, Poledna S, Hatayama M, Kumagai K (2017) Evaluation of evacuation plan: taking account of responsiveness of performance to the uncertainty around hazards and the compliance of evacuees. In: Proceedings of the 15th international conference on computers in urban planning and urban management, Adelaide, Australia, Conference USB (ThuA62_35042)

Rowan KE, Botan CH, Kreps GL, Samoilenko S, Farnsworth K (2010) Risk communication education for local emergency managers: using the CAUSE model for research, education, and outreach. In: (Robert L. Heath and H. Dan O'Hair eds) Handbook of risk and crisis communication. Routledge, pp 359–382

Sun Y, Nakai F, Yamori K, Hatayama M (2017) Tsunami evacuation behavior of coastal residents in Kochi Prefecture during the 2014 Iyonada Earthquake. Nat Hazards 85(1):283–299

Gaming Approach to Disaster Risk Communication: Development and Application of the "Crossroad Game"

Fuhsing Lee and Katsuya Yamori

Abstract Recently, empowerment of local residents in disaster-affected areas has been encouraged during the disaster prevention and recovery period. However, local residents still tend to define disaster prevention and recovery work as the responsibility of experts and the government. From a risk communication perspective, these problems are the result of the traditional framing of disaster education as a one-way communication process. In this study, we attempt to overcome the problem of one-way communication. To overcome this problem, we used an interactive communication tool called "Crossroad," a game to assist people in thinking about a disaster. We first investigate the challenges of risk communication. Next, we introduce the concept and rules of the "Crossroad" game. We also report action research for developing a bottom-up strategy in which local people create their own "Crossroad: Oarai" questions in a community called Oarai Town, Ibaraki Prefecture in Japan. We found that local people from the community composed their own questions, rather than simply answering those from the government, mass media, or others from the outside, which demonstrated that they have the confidence to face such risks alone.

Keywords Empowerment · Risk communication · Crossroad · Bottom-up · Recovery

1 Introduction

Recently, empowerment of local residents in disaster-affected areas has been encouraged during the disaster prevention and recovery period. However, local residents still tend to define disaster prevention and the recovery work as the responsibility of experts and the government. After the occurrence of a major disaster, people tend to ask government officials and experts to take full responsibility for the rescue and recovery process in disaster-affected areas. At the same time, people expect the experts to teach them the knowledge that will allow them to avoid the same situation

F. Lee (✉) · K. Yamori
Kyoto University, Kyoto, Japan
e-mail: franaloha@drs.dpri.kyoto-u.ac.jp

K. Yamori (ed.), *Disaster Risk Communication*, Integrated Disaster Risk Management,
https://doi.org/10.1007/978-981-13-2318-8_4

in the future. However, as time passes, people forget about disaster prevention, until the next disaster strikes.

There is another problem. The knowledge taught by experts is not suitable for all situations and communities. For example, in regard to the question, "What should we do after an earthquake occurs?", there are many answers, the most popular being "Drop, cover, and hold on." However, it must be asked whether this is the best answer to help people face the situation in all cases. If someone is mountain climbing or playing on the beach, how can they "Drop, cover, and hold on"?

It is clear that there is more than one answer regarding what to do in disaster situations. Nowadays, disaster management textbooks and manuals are growing larger and larger because they provide different answers for different situations. However, it is impossible to describe all scenarios in one book, and actual events and scenarios are often quite different from those described in books. As a result, the contents of textbooks and manuals do not include up-to-date information regarding the more frequent disasters and serious situations. This amount of complicated knowledge regarding disaster prevention can cause confusion, and experts, government officials, and people tend to blame each other for this confusion. For instance, people often think experts and officials should provide precise and straightforward disaster prevention strategies, while experts and officials frequently criticize people for lacking awareness and knowledge of disasters.

The problems also happened in the recovery process. People in the affected areas are regarded as helpless and ignorant. Experts and government officials utilize their specialized degrees to identify the degree to which a disaster has affected an area and what type of assistance the area needs; however, they can tend to ignore what is needed by the residents. For example, the senders tell the receivers in a top-down manner that "We have already designed a great plan for you and it is the best way." In fact, these plans always differ from the local people's expectations.

The problem is not how to value the precision of knowledge or how to complete the needs and expectations for affected areas, but rather how to ensure good communication between experts and people. From the risk communication perspective, these problems are the result of the tradition to frame disaster education as a one-way communication process. One-way communication represents a form of top-down authority in which the sender delivers a message to a receiver, but no information flows back. Under this type of disaster prevention education, the experts and governments who develop new technologies and education are the senders, and local communities who receive this knowledge are the receivers. This form of communication causes experts and governments to undertake more assignments.

At the same time, people develop a reliance on the senders and lose the ability to face disasters on their own.

In this study, we try to find a way to overcome the problem of one-way communication. To overcome this problem, we used an interaction communication tool to assist people in thinking about a disaster. After the Great Hanshin (Kobe) earthquake in 1995, the Japanese people discovered that one-way communication for disaster education was not effective and decided that it was time to practice bi-directional communication. The Disaster Prevention Research Institute at Kyoto University developed a disaster prevention game called "Crossroad" (Yamori 2007). "Crossroad" is based on the idea that a gaming approach could make a positive impact on disaster risk communication. Experts, governments, locals, and other stakeholders could be the players in this game, and they could discuss a variety of issues with and learn from each other.

The participants who play the "Crossroad" game do not only receive knowledge from experts or government officials, but also are able to express their considerations freely. Playing the game "Crossroad" gives participants the ability to overcome the problem of one-way communication.

In this paper, we first investigate the challenges of risk communication. Next, we introduce the concept and rules of "Crossroad." To reverse the top-down hierarchy, we report action research for developing a bottom-up strategy in which local people create their own "Crossroad: Oarai" questions in a community called Oarai Town, Ibaraki Prefecture in Japan.

2 Challenges of Risk Communication

To change the one-way relationship between experts and citizens, the following three challenges must be addressed.

2.1 Risk Communication and Experts

"Risk communication" is currently associated with a number of varying concepts, suggesting that the role of experts in society is changing. Yamori (2007, 2009) identified two risk concepts in regard to when people face disasters: "neutral risk" management and "active risk" management.

"Neutral risk" management means risk in a narrow sense, judging risk based on the level of severity, for example, the probability of death in an aviation or natural disaster. However, experts calculate a numerical value from the natural environment and society without considering how local people perceive these risks.

"Active risk" management means risk in a broad sense, focusing on how risk invokes positive human behavior, as opposed to statistics. In this case, risk is not a neutral condition; instead, it relates to how people face a disaster and take action. People can attempt to understand information provided by experts, and they also have the authority to work with these experts. For example, people can work with experts by volunteering to help rescue survivors in disaster-affected areas.

The positions of experts are changing from communicating knowledge to other people ("neutral risk") to facing risks with others ("active risk"). Here, "active risk" refers to situations in which experts play the role of risk "partners," meaning that they are not positioned at the top of the teaching hierarchy; rather, they work in parallel with people to solve problems. Regarding earthquake prediction, this means that they are concerned with not only the science behind earthquakes, but also how it impacts society. Sometimes, predictions from experts spur arguments within the community. For example, "If the prediction is wrong, why do we need to spend so much money on prediction methods?" and "Should we spend so much tax revenue to strengthen earthquake resistance in schools?" To address these questions, responsibility should be taken not only by governments and experts, but also by all the citizens and stakeholders involved in the decision-making.

2.2 Balance Between Self-help, Mutual Assistance, and Public Assistance

The second challenge is to find the most appropriate balance between self-help, mutual assistance, and public assistance. From the Great Hanshin Earthquake of 1995, Chuetsu Earthquake of 2004, and Tōhoku Earthquake of 2011, people learned the lesson that to save lives, they cannot rely solely on public assistance or support from governments and experts; mutual assistance from individuals and communities is also needed.

However, a challenge remains in that although public assistance can help to develop knowledge and technologies, nobody seems to care about how to develop such tools for self-help and mutual assistance. Therefore, people still consider public assistance to be reliable, and experts feel as though they have a strong responsibility to develop knowledge and new technologies. This type of one-way communication (experts ask people to enhance their ability of self-help and mutual assistance, but continue to increasingly strengthen public assistance) then forms a communication loop.

It is time to break this loop. To do so, tools or methods for self-help and mutual assistance need to be developed. It also may be possible to develop alternative strategies that challenge the "educator versus learner" and "supporter versus receiver" hierarchies. By incorporating self-help and mutual assistance into society, people can transform from students into teachers, and experts from teachers to students.

2.3 The Next Generation

The third challenge is how to improve disaster education and the ability of communities. Preparation for disaster prevention can be divided into three periods. First, "one year," which refers to preparation for disasters that occur during a 1-year period, such as typhoons. Japanese carry out prevention drills for such disasters once a year. Second, "10 years," which refers to disasters that could occur over the next 10 years, which people find more difficult to prepare for. The social insurance system protects against this type of risk. Third, "100 years," which refers to disasters such as large earthquakes and tsunamis that occur once every 100 years. It is difficult to extend knowledge regarding disaster prevention to subsequent generations beyond the third period.

Some information can be transferred to the next generation, but people find it difficult to remember and continue discussing such plans after more than 50 years. Storytellers cannot hand down reliable information past "100 years." Therefore, disaster prevention for "100 years" needs to take the community, society, technology, knowledge, experiences, and culture into account.

It is difficult to solve these three challenges in the short term, so a large number of approaches for repeating these practices are needed. We found that games are a helpful tool for overcoming those problems.

As stressed by Duke (1974), gaming is a very effective context for realizing what are referred to as "multilogues," a variety of narrative interactions, such as persuasion and negotiation, that occur naturally among game players. Thus, a game setting affords the opportunity for multiple narratives among participants. Gaming also represents the recognitions and conflicts of individual stakeholders.

As Duke (1974) mentioned, games/simulations are "the future language," a new form of communication that will allow citizens, scholars, policy-makers, volunteers, and reporters in the mass media to formulate and consider a variety of scenarios. Gaming is not a language for describing everything objectively, such as science articles, but rather a language for allowing everything to be described subjectively, from the perspective of each individual. Everyone can use the practical language of gaming to achieve the aim of "active risk."

The disaster prevention game "Crossroad" is one such approach in Japan. "Crossroad" is a game that offers a space in which the players, equally balanced between experts and locals, can accumulate knowledge about self-help and mutual assistance and pass it on to subsequent generations.

3 "Crossroad" Game

3.1 "Crossroad: Kobe"

In 2002, as part of the Ministry of Education, Culture, Sports, Science and Technology's Special Project for Earthquake Disaster Mitigation in Urban Areas, researchers interviewed Kobe City officials who had experienced the Great Hanshin Earthquake of 1995. Since then, a huge volume of interview records has been accumulated. Based on the accumulated interviews, the "Crossroad" game was developed by Kyoto University and the "Crossroad" team. The researchers discovered that stories from these interviews described existing dilemmas. Based on these dilemmas, they designed a card game (Kikawa et al. 2004), which was completed in 2005.

The rules of "Crossroad" are simple. About five or seven players sit around a table, each holding a "Yes" and a "No" card. One player reads the text written on a problem card, and the rest of the players show either the "Yes" or "No" to indicate their answer. They then have to give their own reason regarding why they showed the "Yes" or "No" card to the other players. If a player shows the same card as the majority of the other players, they get 1 point. When only one player shows a "Yes" (or a "No") card, they get 5 points. The purpose of the game is to deepen discussions on why players come to think as they do, instead of dismissing minority opinions (Kikawa et al. 2004).

The original version of this game is "Crossroad: Kobe." The questions were based on the Great Hanshin Earthquake of 1995. For example, one question is "(Imagine that) you are a city official in charge of food distribution. A few hours have passed since an earthquake struck the area. You obtain information that 3000 people have been evacuated to a shelter. At present, meals for 2000 people are secured for the shelter. Whether additional meals will be obtained is unclear at present. In this situation, will you distribute the meals to the evacuees right now? Yes or No?" If you choose to distribute the meals to the evacuees, then there would be 1000 people complaining that this is unfair. If you choose not to distribute the meals, then they will be useless. Another question is "You are transferring patients to other hospitals. A TV cameraman is recording video of them. You find the video-recording hard to endure. Do you allow the video-recording to continue? Yes or No?" If you choose "Yes," then you have to worry about the privacy of the patients. If you choose "No," then an important means of recording this disaster might be lost. After considering these questions, you discover that there is no model answer.

Although the questions from "Crossroad: Kobe" focus on disasters, the format of the game is simple, so experts and researchers have developed the game to address other issues, such as child-rearing, the environment, food safety, casework, and themes other than disaster prevention. A common feature of "Crossroad" is that all the versions have been produced by experts or governments, even though all the questions are based on actual experiences from local people.

"Crossroad: Kobe" provides a universal framework for other participants to simulate a disaster. Through this game, people can learn knowledge from each participant and accumulate self-help mutual assistance experiences. To date, "Crossroad" has been played for more than 10 years for disaster prevention education. This game has shown the ability to overcome the three challenges required to address the issue of one-way communication, as described above. However, one remaining problem is that experts made the "Crossroad" questions as the "senders, while local people only play the role of the "receiver."

In this paper, we describe our attempts to change the position of "sender" and "receiver" by asking that the people involved be given a chance to create their own version of "Crossroad," namely, "Crossroad: Oarai." "Crossroad: Oarai" is a version of the game produced by people living in Oarai Town in Ibaraki Prefecture.

3.2 *"Crossroad: Oarai"*

The town of Oarai has 7000 households and a population of 18,000. Located at the approximate midpoint of Ibaraki's Pacific coastline, the town forms the eastern end of Higashi Ibaraki District (Fig. 1). The town's key industries, fishing, and seaside tourism make use of the region's abundant marine resources. In fact, before the earthquake, Oarai attracted the highest number of tourists in the prefecture.

When the 2011 Tōhoku Earthquake hits, 10% of the town was inundated by a tsunami measuring up to 4.7 m in height. One person was killed by the earthquake, but nobody was killed by the tsunami. The physical recovery process in the town has proceeded smoothly, but the community continues to experience concerns over radiation from the Fukushima Daiichi Nuclear Power Plant accident, as well as from *fuhyo higai*, which refers to harmful rumors and misinformation regarding the nuclear accident. The fishery and tourism industries in Oarai are in operation for only half of the year. After diverse efforts by locals, these industries have improved. The most effective efforts to date have involved a series of projects linked to a televised anime series.

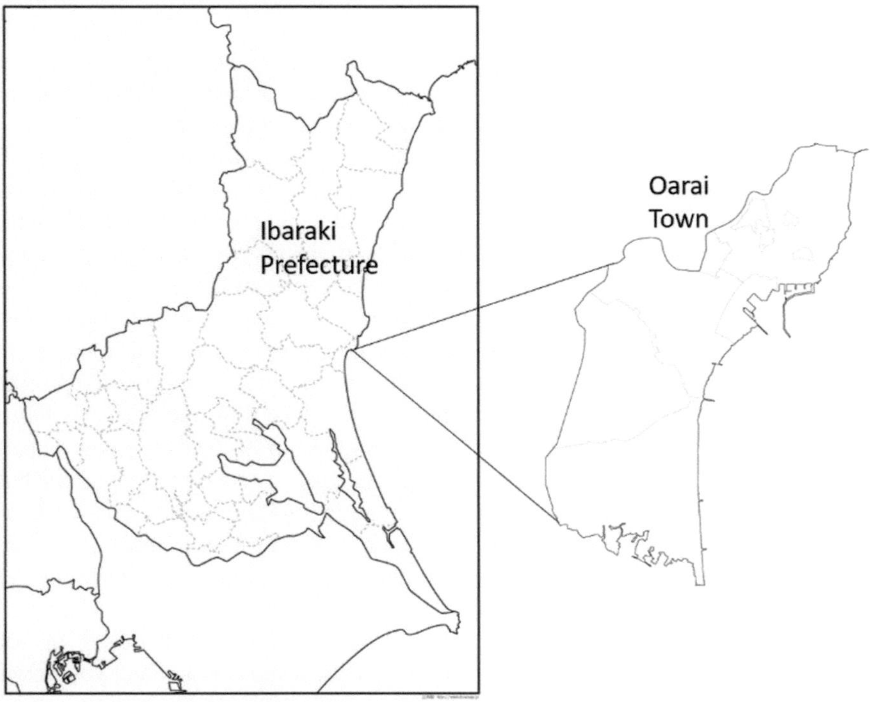

Fig. 1 The map of Oarai Town, Ibaraki Prefecture. Reprinted from Lee et al. (2019). Copyright 2019

After November 2012, I (first author) conducted participant observation and interviews in Oarai Town involving volunteer firefighters, fishers, elementary school students, shop owners, and hotel owners (Lee et al. 2015). I found that they experienced conflict when talking about the disaster. However, people did not have the chance to record their important stories for subsequent generations. I also found that, in the recovery process, people felt that the government did not really understand what the people needed. At the same time, people were also disappointed with the mass media because reporters tended to have a negative image of Oarai Town, for example, giving the impression that it was contaminated with radiation, or overrun with garbage, pity, and hopelessness.

We (both authors) held a "Crossroad: Kobe" workshop on March 29, 2013 and we found that "Crossroad" could help Oarai Town record their disaster experiences and help to document the relationship between people, the government, and the mass media. In March 2014, I attended an Oarai Town disaster prevention education meeting and promoted a "Crossroad" project involving having local residents create "Crossroad: Oarai." After I obtained consent from the local government, I started this project.

We started to produce "Crossroad: Oarai" from November 2014. I explained the meaning of the "Crossroad" game to people I knew in Oarai and asked them to formulate their own questions.

In contrast to the previous "Crossroad" questions and knowledge produced by experts, the "Crossroad: Oarai" project asked people who were affected by the actual disaster to explain their stories and conflicts in their own words.

"Crossroad: Oarai" provides a local framework that allows residents of the Oarai community to continue discussing localized questions. The theme of the questions could be divided into three types: evacuations after an earthquake and tsunami, radioactive fallout, and industrial recovery. The participants composed 25 questions.

For example, regarding evacuation from an earthquake and tsunami, if you were an 80-year-old man and a tsunami was approaching, would you accept an offer to evacuate in your friend's car (Yes) or run to higher ground on your own (No)? Likewise, if you were a fisher and a tsunami was approaching, would you first protect your boat (Yes) or take care of your family (No)? Regarding radioactive fallout, imagine that you own a tourism company and your town has suffered problems related to radiation. You wish to hold an event to promote tourism and invited reporters, but you realize that the reporters will focus on the radiation issue, would you let them come (Yes) or not (No)? As another example, imagine you are a resident of an area affected by radiation after a huge disaster and your community now has no food to eat. Your neighbor offers you some fish that he/she could not sell, but assures you that they are safe and free of radiation. Would you eat the fish (Yes) or not (No)? Regarding industrial recovery, imagine you are a shop owner, and your town is developing tourism related to anime. Although you do not know much about anime, would you promote and sell anime products in your shop (Yes) or not (No)?

We video-recorded locals as they told their stories and saved 15 video clips on a DVD. I also asked people to design question cards to show the features of their community. They wrote their own questions on a postcard and then drew illustrations. The cards shown in Fig. 2 were designed by the participants.

I held a workshop for "Crossroad: Oarai" in Oarai Town eight times. We first played the game using the already completed questions. Next, I asked the participants to make their own questions. In the workshops, the people of Oarai Town discussed their opinions with others to plan the next step for development of the community (see Table 1).

Below, we cite the questions in two cases to examine whether "Crossroad: Oarai" achieved the goal of breaking the mode of one-way communication and resolving the three challenges introduced earlier.

Fig. 2 Cards designed for use in "Crossroad: Oarai." Reprinted from Lee et al. (2019). Copyright 2019

Table 1 Interview list. Modified from Lee et al. (2019). Copyright 2019

	Sex	Occupation	Time
A	M	Volunteer firefighter	2014–2016
B	M	Lifesaver	July 8, 2014
C	M	Fisher	2014–2016
D	M	Fisher	2014–2016
E	M	Hotel owner	November 27, 2014

Q1. You are a volunteer firefighter. After a big earthquake occurs, a tsunami warning alarm is sounded. You instruct cars on the street just next to the coast be evacuated. You know the tsunami will come after 20 min. Would you run away or stay there for your work? Choose "Yes" to run or "No" to stay.

This question was composed by volunteer firefighter A. The first time that I met A in December 2012. He told me this story at that time. He still remembers this experience clearly.

Thus, volunteer firefighter A stayed on the coastline at that time. After the 2011 Tōhoku Earthquake, he knew that many firefighters in the Tōhoku Region had died because they did their work without prioritizing their own safety. Therefore, he changed his views and decided to take the responsibility of leading others to safety, which could save both his life and the lives of others. He created this question and expressed the following emotions in the video clip from "Crossroad: Oarai":

> I had no idea that the tsunami would be so strong. I directed the evacuated cars on the street just next to the coast. The water had come up to us. I did not think that the water would come any higher, so I continued my work. After that, I thought that maybe I should not stay there. If the tsunami had been as bad in our town as it had been in Minamisanriku, then we all would have died. Being volunteer firefighters, we should seek safety first, and then let people see what we were doing so that they could follow.

However, there were some opposite opinions. Among the players who answered "Yes" to run, some of the reasons were "It is important to save the lives of others, but I have to save my own life first"; "I can lead them in a safer direction rather than stay in a dangerous area"; and "If you had asked me this question before the 2011 Tōhoku Earthquake, I would have said 'No, I will stay to save others', but since that disaster, I know I have to protect my own life first." The reasons given by some of the players who answered "No" to stay were "This is my responsibility as a volunteer firefighter" and "I will stay for a while, but if I feel something is wrong, I will evacuate as soon as possible."

A leader of lifesaving team in Oarai, B, noted that "I agree with A, and the law also tells us that we should prioritize saving our own life first. However, if a tsunami will come after 20 min, and there are still children at the beach, I do not think I have the courage to run instead of saving the children." Another volunteer firefighter from Miyagi Prefecture said, "A expressed an opinion that we aren't brave enough to put into words."

The aim of this question is not to argue about whether the rescue team or firefighters should evacuate first, but rather to help people understand that the rescue teams have their own conflicts and difficulties. Firefighters wish they could prevent deaths, while at the same time, they also expect people to recognize that they are not gods or superheroes and that they have limitations when carrying out their mission. People also should improve their self-help capabilities.

The next question is as follows:

Q2. You are a fisher from an area affected by the nuclear crisis, and thus your fish cannot be sold. You want to post an article on your Facebook page to express your sadness, but you also worry about being misunderstood by Internet users. Do you post your article? Yes or No?

This question was composed by C, a leader of young fishers in Oarai Town. C and his fishing companions were members of the fisher association. C posted a picture of himself catching a large Suzuki fish on his Facebook page on November 9, 2012. He commented, "Even though I caught this large Suzuki fish, I cannot sell it (>.<)." Suzuki is a type of fish that was among the best selling for Oarai's fishery. After the 2011 Tōhoku Earthquake, radiation levels were higher than normal, so Suzuki became a restricted product. After C posted the picture, his friends left some comments for him on Facebook. Some said they felt pity for him because of the disaster, and others encouraged him not to be upset. C regretted posting the picture. He expressed his feelings as follows in the video clip:

> I posted a picture of myself catching a big fish that I couldn't sell because of the radioactive contamination problem. I regret this. I just felt that it was such a pity to throw the fish away, and I was sad that I could not provide delicious fish to people. It was not because the fish was contaminated, but because I was misunderstood by the people who read my comments. People thought my sadness was from the nuclear contamination. They felt sorry about this, but they still did not want to buy my fish and did not trust us. I did not want to be misunderstood; therefore, I regretted posting that picture and writing those comments.

Radioactive fallout was the most serious problem affecting Oarai Town. Not only fishers but also other workers had similar experiences. Among the players who answered "Yes," some of the reasons were "The news would not report on Oarai, so we have to promote it ourselves."; "If some people do not believe us, then there must be some people who do."; and "I will post information after the arguments regarding the radioactive fallout have subsided." Among the players who answered "No," some of the reasons were: "It is too difficult to dispel the rumors; if I post something that is not true, then people on the outside would misunderstand that forever." and "I cannot control the Internet, so I should be careful if I publish information."

I held a "Crossroad: Oarai" workshop at the Oarai Fishery Association on May 23, 2014. Seven fishers attended the workshop to play "Crossroad: Oarai" and create their own Crossroad questions.

In this workshop, the fishers played with the question composed by C. Six fishers chose the answer "Yes" (post the information), and only one chose "No" (do not post the information). The only person who selected "No" was C. I played C's video clip for all the participants. C also expressed his opinion to other fishers; one of them D told C, "You think too much." D also stated "I think we should post the right thing we believe on my Facebook page." C said, "After the 2011 Tōhoku Earthquake, we went to Tsukiji, Tokyo to promote Oarai's fish products. A customer asked me, 'Even though you showed the food safety certification, can you guarantee that the fish is really safe from radiation?' I felt very upset with this question because I could not answer this customer well." Other fishers said they had similar memories. After they exchanged their memories, they started to discuss whether they should maintain their fishing career, relying on certification by the government, and what they could to do on their own. They considered that they could not explain and totally understand the meaning of the government certifications. They mentioned that they could use their creativity, for example, by using Facebook, posting fish recipes and pictures of themselves, and showing their daily lives to society. They hoped that by doing this, people would understand their situation better and like fish more.

After the workshop finished, C told me the following:

> When everyone told me that I thought too much, I felt relief. I have been worried about this stuff for a long time. However, nobody seemed to care. At the same time, I am also surprised that we worked together over the past 10 years. I thought we knew each other deeply and that our ideas were the same. However, after this workshop, I discovered that we had many different ideas.

In relation to the consensus they reached, C said:

> Even though I only could choose one answer for the Crossroad question, I found that we could mix opinion A and opinion B together to create opinion C, and this was our final answer. The point is, people are not forced to accept an answer. The answer is reached by gathering all the opinions based on discussions and decisions.

E, who owns a hotel, created his own questions for "Crossroad: Oarai." E said: "*At first, you (the first author) asked me to create and write down my questions. I played this game and I read other questions, and then I started to think, if we resolve all the questions in "Crossroad: Oarai," then our town would be a perfect community.*" This case showed that through this game E started to consider disaster prevention and recovery work on his own rather than looking to the government, as he had done before.

4 Discussion

In this study, we tried to address the challenges of disaster education, namely, how "Crossroad" changed the relationship between experts and local people, and how "Crossroad: Oarai" helped inspire self-help and mutual assistance in individuals and the community. Based on the results, we expect this to be a long-term project passed on to subsequent generations.

In "Crossroad: Oarai," the role of experts, including researchers (authors), government officials, and the mass media, is not as "senders" but as "partners." Experts did not force people to state the correct answer or learn the appropriate knowledge; they only listened and expressed their opinions. The main role of people in Oarai Town was to create the contents of "Crossroad." The people who participated in creating "Crossroad: Oarai" exchanged their experiences of suffering through the earthquake, tsunami, and radioactive fallout to create questions for the game. They also expressed their feelings about the real disaster while playing the game and in video clips. They defined what problems they faced, and how they could solve them, on their own.

We mentioned earlier that the aim of disaster prevention games is to gain not only scientific knowledge, but also knowledge from creating and playing the game. The method of creating and playing the game could ease the intensity of participating and help further the aim of avoiding risk. Through the process of developing "Crossroad: Oarai," people who discussed the problems that were difficult to resolve in the recovery process discovered answers beyond a simple "Yes" or "No."

Next, we investigated how well "Crossroad: Oarai" addressed the challenges of self-help and mutual assistance. In "Crossroad: Oarai," people discussed their own problems and developed knowledge and various methods for the future. The point is that they attempted to resolve such problems by themselves rather than asking experts or government officials, as they had done before. In the case of volunteer firefighter A, he described his conflict and emphasized that he would evacuate first in the video clip. He also conveyed an important message to people in the community that firefighters are not superheroes, so people should evacuate as soon as possible on their own.

Through the case of fishers, fisherman C regretted the negativity brought to the fishing industry caused by his picture. After playing the game, his company helped to resolve his conflict. They also started to discuss the next step together. The fishers said they would continue posting information about the fishing industry, but refrain from remarking on radiation certification, which they did not understand. In addition to individuals, this process is also important to the community. Through this process, people could reconstruct a consensus for the direction of development and recovery of the industry and formulate an approach regarding how to obtain support from outside. This process reversed the idea that the government, mass media, and outside experts should assist the community, as had been done in the past.

To achieve the goal of disaster prevention for "100 years," "Crossroad: Oarai" helped people to record and develop a disaster prevention culture in the local community. Through the process of creating "Crossroad: Oarai," we found that players avoided conflicts stemming from seeking the right or wrong answer and ultimately created new ideas for the future. They discovered the problems by themselves and started to think about how to make their community better, as stated by E.

5 Conclusion

In this paper, we introduced the process of developing a tool, "Crossroad: Oarai," for resolving problems with one-way communication and challenges with disaster prevention. "Crossroad: Oarai" accumulated serious and unresolved problems in the community, meaning that Oarai has hazards and risks to deal with. The key point is that local people from the community composed their own questions, rather than simply answering those from the government, mass media, or others from the outside, which demonstrated that they have the confidence to face such risks alone. As mentioned by the local people, each question in "Crossroad: Oarai" represented a unique opinion for the improvement of the community. If more discussions are held and activities are carried out, a better Oarai community could be achieved in the future.

References

Duke R (1974) Gaming: the future's language. Sage Publications, New York

Lee F, Miyamoto T, Kondo S, Yamori K (2015) The disaster recovery process and rashomon effect: the case of Oarai Town, Ibaraki. Jpn Assoc Qual Psychol 14:38–54 (Japanese)

Lee F, Miyamoto T, Yamori K (2019) Empowering local residents during disaster revitalization via self-supported study: creating the disaster game "Crossroad: Oarai". Jpn J Exp Soc Psychol 58(2):81–94 (Japanese)

Kikkawa T, Yamori K, Ajiro T, Hayashi H (2004) "Crossroad: Kobe": a training tool for disaster preparedness and response. In: Kriz WC, Eberle T (eds) Bridging the gap: transforming knowledge into action through gaming and simulation. SAGSAGA, Munich, Germany, pp 245–253

Yamori K (2007) Disaster risk sense in Japan and gaming approach to risk communication. Int J Mass Emerg Disasters 25(2):101–131

Yamori K (2009) Action research on disaster reduction education: building a "community of practice" through a gaming approach. J Nat Disaster Sci 30(2):83–96

Collaborative Community Weather Information

Kensuke Takenouchi

Abstract Weather and hydrological information have rapidly improved with developments in science and technology, but community engagement with disaster prevention has been declining gradually. The relationship between available information and community engagement can be separated into issues of an overdependence on information or an attitude of waiting for more information by recipients of weather and hydrological information. To resolve these issues and construct a new relationship between the providers of information and users, we discuss collaborative community weather information (CCWI), a new approach to weather and hydrological information created by promoting risk communication among various stakeholders, including residents. Since July 2014, the author has been undertaking practical research on CCWI in the Nakajima School District in Japan. We consider the effects and future vision of CCWI through a case study in which three CCWI systems were created through collaboration with the Nakajima local government, weather office, river manager, community council, residents, and researchers. We have observed the operation of the Nakajima CCWI system and report our findings in this chapter. We conclude that CCWI can facilitate better risk communication in disasters as collaboration between providers of meteorological information and local recipients of risk information.

Keywords Weather information · Hydrological information · Risk communication · Community culture in disaster prevention · Community collaboration

1 Introduction

> Don't cross the river when you hear thunder in the morning. (Japan)

This is a famous saying in Japan concerning disaster prevention. Thunder in the morning means the possibility of heavy rainfall with an active front, which can lead

K. Takenouchi (✉)
Disaster Prevention Research Institute, Kyoto University, Kyoto, Japan
e-mail: takenouchi.kensuke.3x@kyoto-u.ac.jp

K. Yamori (ed.), *Disaster Risk Communication*, Integrated Disaster Risk Management,
https://doi.org/10.1007/978-981-13-2318-8_5

to a flood. The topography of Japan is formed by the precipitous mountains of the Ring of Fire, and rain falling in an upper area reaches lower areas within several hours. Therefore, hearing thunder in the morning makes us prepare for heavy rainfall or floods.

All over the world, there are proverbs and lore about predicting bad weather (and consequently acting to mitigate its dangers):

> March comes in like a lion, and goes out like a lamb. (United Kingdom)
>
> April showers bring May flowers. (United States, etc.)
>
> A pig pulls straw before heavy rainfall. (Russia)
>
> Swallows fly low, snakes appear on paths, chickens don't come back to roost; soon will come heavy rain. (China)
>
> Whales in the canal, you will have a storm. (Portuguese)

Such lore has been passed down from generation to generation as a means of avoiding disasters and much of it is still remembered in each local area, although opportunities to hear such proverbs and lore are becoming rarer. However, science and technology have increased the amount of available information about weather and hydrological conditions, and receiving such information via television or smartphone is very convenient and commonplace. As a result, the public's perception of risk has become much more passive, instead of the active local awareness implied by the proverbs and the lore. In this chapter, we will consider whether a more community-based risk awareness can be fostered by the collaboration between developed information and a variety of local measures.

As the traditional warnings imply, human beings have historically always faced natural disasters. Figure 1 shows the numbers of deaths worldwide from each natural disaster type for the period 1900–2017 based on the EM-DAT database. The cases extracted are climatological, geophysical, hydrological, and meteorological disasters in the Natural Disaster category. In Southern Europe, Western Asia, Southeastern Asia, South America, Central America, and the Caribbean, earthquakes and tsunamis cause the most deaths. In Australia and New Zealand, Western Europe, the Russian Federation, Eastern Europe, Southern Asia, and Africa, extreme temperatures and drought are the biggest killers. The focus of this chapter is meteorological disasters, like floods, landslides, and storms. Southeastern Asia, Eastern Asia, Central Asia, Northern America, and Central America have high death rates from this type of disaster. As this data shows, each region has its own characteristic disaster risks. However, these are just statistical results, and more localized regions could have different risk profiles.

Focusing on recent years, the World Disasters Report 2015 (IFRC 2015) reported that the number of natural disasters between 2005 and 2014 was 3809 and the number of deaths was 764,204; the death rate in Asia was especially high. Among these disasters, the number of floods was 1751 and the number of storms was 988; so water-related disasters occurred most frequently.

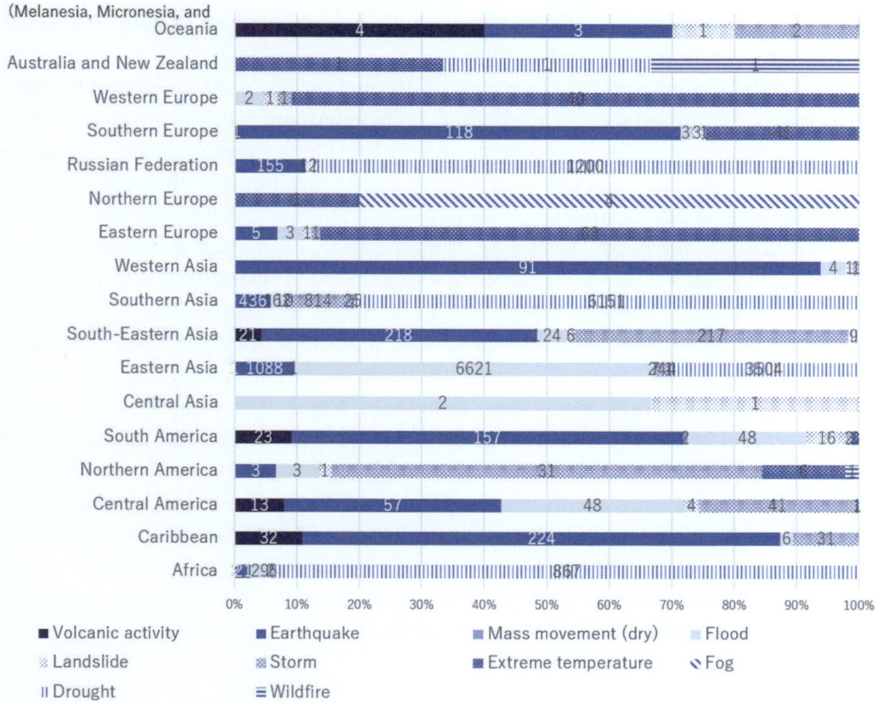

Fig. 1 Number of deaths worldwide from each natural disaster type for the period 1900–2017 (unit: thousands) *This figure was created by the author based on EM-DAT data (EM-DAT 2018) for climatological, geophysical, hydrological, and meteorological disasters in the Natural Disaster category. Causes leading to less than one thousand deaths are omitted

Human beings have undertaken disaster prevention measures for as long as societies have existed. The Sadd el-Kafara dam, which is regarded as the oldest dam, is thought to have been constructed around 2750 B.C., so disaster prevention measures have a long history. In general, disaster prevention measures can be broadly separated into structural measures and nonstructural measures. In addition to developing structural measures, nonstructural measures, such as disaster prevention drills in organizations and communities or services of risk information based on meteorology or hydrology, are becoming increasingly relied upon. However, some of these nonstructural measures do not work efficiently and their effects can weaken over the long term. One such negative effect occurs when disaster information, such as weather and hydrological information, is provided. Some examples of this issue are found in Japan. At present, an overdependence on information or an attitude of waiting for more information is the social issue in disaster prevention. With the advancement and greater sophistication of weather and hydrological information, residents lose the attitude of self-help and become more dependent on outside professionals. Of course, this does not mean that disaster information is not useful for disaster prevention. However, the original purposes, such as the improvement of residents' consciousness of disaster treatments and an increase of proper disaster actions, have not been achieved enough.

The Sendai Framework for Disaster Risk Reduction 2015–2030 (United Nations 2015) noted the importance of traditional knowledge and local and specific characteristics, highlighting the importance of local residents' participation in disaster mitigation and the need for consideration of each community's characteristics. In one of the framework's priorities, "Strengthening disaster risk governance to manage disaster risk," the need "To assign, as appropriate, clear roles and tasks to community representatives within disaster risk management institutions and processes and decision-making through relevant legal frameworks" was raised. This discussion shows the paradoxical situation that exists in societies that use advanced disaster information. Although sophisticated information can provide timely and proper warnings, it cannot completely prevent damage arising from a disaster.

Weather and hydrological information has been rapidly improving with developments in science and technology, but community engagement with disaster prevention has been declining gradually. In this chapter, we discuss the paradoxical relationship between weather and hydrological information and community culture through a real case study of collaborative community weather information (CCWI), a new approach to weather and hydrological information created by risk communication among various stakeholders, including residents.

2 Present Status of Residents' Participation in the Weather and Hydrological Information Process

Practical research on residents' participation in water-related disaster policies has been mainly conducted in terms of "Participatory Processes" and "Community-based Approaches" in research fields such as "Decision Making for Disaster Treatment," "Early Warning Systems," and "Flood Risk Management."

Figure 2 outlines in four phases the relationship between the technical level of weather and hydrological information and residents' awareness of participatory information. Overall, the technical level of information rises gradually, but the awareness level rises and falls. In the following, each phase is discussed in turn.

Phase I

Phase I is the stage prior to the utilization of weather and hydrological information based on science and technology. In this phase, communities prepare against disasters using their own methods and there is no concept of external sources of weather and hydrological information. The lore discussed in Sect. 1 is typical of community efforts in this phase. The development of technology, such as observation devices, information networks, and analysis models, proceeds separately. In this phase, technological and community efforts have no direct relationship and each sector advances its own approach to disaster prevention.

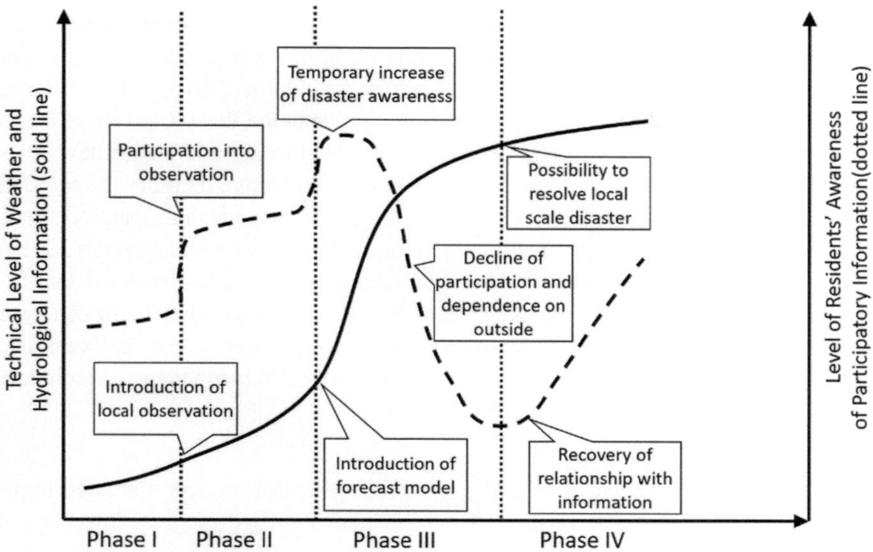

Fig. 2 Outline of relationship between the technical level of weather and hydrological information and residents' awareness of participatory information

Phase II

Phase II is the stage of creating the relationship between technological sources of weather and hydrological information and a community. The representative case is the beginning of meteorological observation in a community. The government starts to use the observation and collection system for rainfall amount, river water level, and so on, and to produce weather and hydrological information based on the observed data. At this time, local residents can participate in the system and can sometimes cooperate by making observations. In this phase, local residents are highly aware of the available information, and if the information is primitive, such as rainfall amount or river water level, residents can participate in its process directly and deeply. Thus, in this phase, it is possible to have good relationship between a community and technological sources of information. Recently, in some Asian and African countries, residents' participation in observation systems has been introduced. Research into the effectiveness of this approach has been supported by the World Bank and other international organizations. For example, some researchers have reported on residents' direct participation in weather and hydrological information around a river basin with support from outside. In these cases, a primitive observation device that can easily be used to observe water levels is used. Smith et al. (2017) point out that exercises using participatory methods combine scientific and community knowledge to help communities gain a better understanding of flood risk and take necessary steps during a flood event. River gauges in Nepal are maintained by the government via part-time staff who, in nearly all cases in Nepal, are members of the local community and are trained to record and monitor river water levels. This results in the situation that, although the government pays for the gauges, information on water levels (in the period immediately leading up to flooding) is provided directly to the local community by one of its members. Gautam and Phaiju (2013) report the case of a community-based approach to early flood warning in the West Rapti River Basin of Nepal. In this case, the preliminary assessment of warning and danger level was carried out by comparing past flood events based on collective memory to recorded data. The communities utilize mobile phones, megaphones, and hand-operated sirens to disseminate flood information at the local level. Macherera and Chimbari (2016) review the existing community-based early warning systems with a view to identifying opportunities to develop a similar system for malaria and summarize the different characteristics of national and community early warning systems. In these examples, residents' participation is comparatively easy and the relationship between technical information and communities is relatively good.

Phase III

In Phase III, highly advanced technology, such as forecast models that need high-performance computers, is introduced into weather and hydrological information systems. As a result, the technological level of the information can increase rapidly, and this introduces a "black box" into the information collection and dissemination process. Although residents' awareness may initially be raised temporarily by the introduction of brand-new information, it declines soon afterward. This means that information that residents had a relationship with is externalized to professionals, and this externalization weakens residents' participation. In this phase, residents begin to regard information as a thing to be given and take a passive attitude toward it. That is to say, the increased sophistication can weaken direct engagement with the information, which goes against the original aim of increasing the awareness of disaster risk and the ability to take proper actions for disaster prevention. Of course, if the actions to utilize more advanced and more accurate information are taken in a community, the provision of new information will be more effective, but in real cases, this approach does not work efficiently. For example, at real disaster scenes in Japan, we can find negative effects of information, such as overdependence on information and the attitude of waiting for more information. At present, therefore, we need to reconsider what relationship we need to construct between information and users in societies where highly advanced and highly sophisticated weather and hydrological information is provided.

In this phase, the approach taken is generally to improve the understanding of the information provided in a top-down way rather than to promote a direct relationship with the information as is done in Phase II. Evacuation drills based on the information are very popular. In the field of risk management, there have been some trials to strengthen cooperation among stakeholders through the provision of various types of information. Almoradie et al. (2015) separate the design of a flood disaster management platform into two modules called the collaborative platform (CP) and the collaborative modeling exercise (CME). In the CME, participation occurs via a web-based decision-making environment. The participants evaluate and rank alternatives based on their own preference regarding the importance of the identified objectives. This platform was applied in two real case studies: the Cranbrook catchment (London, UK) and the Alster catchment (Hamburg, Germany). Maskrey et al. (2016) structured the participatory modeling process into three key stages and co-developed their Bayesian network model through activities designed around a sequence of five model development workshops. As an example of the participatory modeling approach, the development and evaluation of a local flood risk intervention model for Hebden Bridge were attempted.

In Phase III, it is becoming difficult for residents to participate directly, so approaches that try to improve the understanding of the information tend to be taken. However, in rare cases, participatory methods are employed, as in Phase II, to utilize local information for forecast models. Chen and Wu (2014) and Sasada et al. (2015) report the system used in Taiwan by "debris flow professional volunteers." After the Chichi Earthquake, the number of areas vulnerable to debris flow rapidly increased, so various nonstructural preventive strategies were introduced. As one of these strategies, residents in communities near debris flow torrents have been recruited since 2005 by the Soil and Water Conservation Bureau as debris flow professional volunteers to assist with real-time rainfall monitoring, disaster notification, and local evacuation tasks. During routine days, debris flow professional volunteers inform residents about the local environment, areas vulnerable to debris flows, and relevant disaster prevention information. When typhoons or torrential rains occur, these professional volunteers cooperate to monitor the local real-time rainfall using simple rain gauges and report their readings using available communication equipment. Apart from providing rainfall data in areas lacking formal rainfall stations, they also assist with disaster notification and advise the residents to evacuate. Catherine et al. (2012) discuss a case in the Philippines. In this case, manual rain gauges were installed in 20 houses of volunteers living within the Bicol River Basin and the volunteers monitor rainfall as part of the overall early warning system. These records by the volunteers are used in simulations and flood bulletins are issued using the model results as guide. The communities' sense of ownership of the system also has an impact on their level of understanding of flooding. These cases exemplify the potential for cooperation between residents and forecast models.

Phase IV

Phase IV is the future vision of new weather and hydrological information. The scale of the information has previously been limited to some large areas, but the development of information and computer technology can gradually allow simulation models to operate at local scale resolution. That is to say, the information can be relevant to the situation of each community, and residents can use weather and hydrological information in their own specific context. In this phase, the scales of the contexts for the technical information and the community are in reasonable correspondence, which facilitates a better relationship between information and community. However, we will need to overcome the passive attitude to information that was constructed in Phase III to arrive at this better relationship for disaster prevention.

In the next section, we discuss the problem of overdependence on information and the attitude of waiting for more information that occurs in Phase III. We also consider the development of weather and hydrological information for progress into Phase IV. In the fourth section, practical research aiming toward Phase IV, which the author is related to, is introduced.

3 To Overcome "Overdependence on Information" and the "Attitude of Waiting for More Information"

In this section, we begin by addressing the issues of passive attitudes, such as overdependence on information and waiting for more information, which are created by the externalization of weather and hydrological information to professionals. These issues have been discussed many times in Japan (Katada 2006; Katada et al. 2005; Takenouchi et al. 2014, etc.). They are related to the consciousness of a relationship with information and are strengthened by the development of outside information.

Overdependence on information means "no information, no action." Yamori (2013) points out the existence of various types of double bind in disaster information in reference to Bateson (1972). Disaster information disseminates the meta-message that "evacuation should occur only after a message (evacuation information from government); in other words, we should take action only after waiting for the message," which fosters an overdependence on information. Moreover, the attitude of waiting for more information means that "we continue to wait for more information forever." As a result of the custom that information is given, users do not have the attitudes required to understand the information and regard the information as just what is given. Consequently, residents do not take action when they are given the information and continually wait for more information. The results of some surveys show such a situation. To the question "When you get some severe weather information, what will you do?", most of the respondents answered "I will wait for more information," even if the information is very severe (Japan Meteorological Agency 2011). As these facts show, advancement and increasing sophistication can add more value to information from the point of view of natural science, but it does not easily lead to improved utilization.

We can find some reasons for the development of passive attitudes to weather and hydrological information. One important factor is the neglect of users' participation in the development of the information: the basis in natural science is emphasized while the social and local aspects are neglected. When the expression of the information is examined, the situation is very clear: objective and numerical expressions to describe natural situations, like "water level: 8 m" or "rainfall strength: 50 mm/h," are used, and this mode expression is different from that of residents who express disaster phenomena in daily life in terms like "it is dangerous around the bridge" or "there is dirty water is flowing from the mountain." That is, most technical information does not focus on the context of residents and does not reflect their locality. What are the reasons for this tendency? The models to develop the information are based on natural science, so the information basically express natural phenomena. When we process weather and hydrological information, substantial infrastructures such as observation devices and computers are necessary, so most systems are operated by governments. As a result, the information is likely to be general and versatile. Therefore, it is difficult to provide information that is relevant for each locality and suitable for each individual.

Changes of social system influence passive attitudes, too. Societies have externalized professional tasks into professionals who have special knowledge and technologies. The tendency is so natural when the efficiency and condition of social systems are taken into consideration, but, as stated above, this can lead to a decline in the community culture of disaster prevention. Thus, the relegation of various kinds of tasks to outside professionals can lead to a black box of disaster information: users do not need to know the details, with the result that the passive attitude of users is enhanced.

For weather and hydrological information, we can see the possibility to change this situation with the advancement of information technology. At present, the resolutions of the advanced analysis and forecast models are from a few kilometers to a few hundred meters, which are approaching community scales. So, recent models are becoming able to deal with community-level phenomena, which was impossible to do directly before. The horizontal resolution of operational short-term forecast models is about a few kilometers: 2 km (LFM, Japan), 1.5 km (UK, United Kingdom), 2.8 km (COSMO-DE, Germany), 2.5 km (HARMONIE/AROME, Spain), and so on (GDPFS 2015). Flood analysis models have advanced to have resolutions from a few hundred meters to a few tens of meters, and can consider land use, house height or channel distribution, and so on. Of course, in some countries, it is not easy to get such detailed data, but others have enough GIS data (e.g., Rainfall–Runoff–Inundation models, such as NILIM [New Integrated Lowland Inundation Model] in Japan). This use of high-resolution models is getting to the point that it can create an environment in which weather and hydrological information can be provided as concrete local information to communities.

In another approach, it is becoming fashionable to use information technology to analyze a lot of data, and Big Data methods or artificial intelligence (AI) are being applied to understand local situations or social actions. Recently, some countries have implemented policies on Big Data, AI, the Internet of Things (IoT), and so on. In Society 5.0 (Government of Japan 2016), disaster prevention is one of the main topics. This stream can connect real-time weather and hydrological information to local information in communities and helps to generate efficient disaster responses.

In a few decades, some long-term forecasts have shown an increase of extremely heavy rainfall in many areas as a result of global warming (IPCC 2012; Takemi et al. 2016, etc.) These forecasts show the limitations of treatments for disaster risk by outside professionals, such as governments, in the future. Therefore, in Japan, the significance of the self-help rule "protection for myself by myself" is now being discussed. If we wait for information, we cannot protect ourselves. In addition, residents are likely to consider disaster response in each community by themselves. We need to reconsider the importance of externalized weather and hydrological information for each community and reconstruct the relationship with that information. This issue will also appear with the development of weather and hydrological information in countries that employ the primitive methods in Phase II for risk governance.

4 Collaborative Community Weather Information

As stated in Sect. 2, we will need to construct a social system (Phase IV) for a society in which weather and hydrological information has been highly developed and residents have become passive. In this section, we consider a new style of, and role for, weather and hydrological information in Phase IV through practical research into CCWI.

CCWI, which is the research target of the author, is created by a collaboration between the highly developed natural science information of professionals and local and traditional information of residents. That is to say, CCWI is created by the transformation of various professional weather and hydrological information into local information familiar to residents through collaboration among various stakeholders. This collaboration can improve the locality of the information and help residents to be aware of local disaster risk and to understand the local situation for early action in response to a local disaster. For this purpose, an important concept of CCWI is risk communication. By translating technical and diverse weather information into CCWI using familiar, plain, and local expressions, CCWI raises residents' awareness of disaster risk and induces early action before a disaster. Moreover, residents can consider the contents of CCWI by themselves, and this consideration will enhance their consciousness of CCWI as belonging to them. Also, a common risk awareness can be developed by sharing CCWI in the community (Takenouchi et al. 2017).

In the development of weather and hydrological information, CCWI takes the role of reconstructing the information as community information in the situation where there is no chance to participate in the construction of the information. CCWI is also a kind of social system. The participation of various stakeholders and a structure connecting each opinion are necessary to construct CCWI. CCWI is a social system that connects various types of knowledge, including professional and highly technological knowledge of weather and hydrological conditions, familiar and real-life knowledge about local conditions, disaster response expertise, and the government's risk governance procedures. In the practical research described in the next section, through workshops by local government, the weather office, the river manager, the community council, residents, and researchers, each participant considered local disaster risk with each taking responsibility and sharing knowledge. Using this approach, we succeeded in the creation of CCWI systems.

What phase does CCWI belong to in Fig. 2? From the view of residents' participation, CCWI goes back to an earlier high level. However, from the view of information technology, CCWI remains at a higher level. The intermediary to create this condition is risk communication.

The development of weather and hydrological information creates the situation that there is little space for users to become involved in the construction of this information. This situation may be creating an overdependence on information for users and preventing them from understanding the relationship between themselves and the disaster risks that the information provider is trying to convey. Improvements in weather and hydrological information, such as refinement of models and increasing image resolution, are conducted mainly by experts and those who take part in the construction of the information. Therefore, improvements have focused on the content of the information rather than the social system comprising information composers and users. In other words, the participation of residents in the construction of weather and hydrological information has not been considered. CCWI is an approach that aims to improve the social usage of weather and hydrological information from the perspective of social relationships. Under this system, information users and composers discuss the information on even ground, and consider the social usage of the information. In other words, it aims to create mutual disaster risk communication between information composers and users through weather and hydrological information and to facilitate co-composition of the information (Takenouchi et al. 2017).

5 Case Study of Collaborative Community Weather Information

In this section, we consider the effect and future vision of CCWI through a real case study. Since July 2014, the author has been undertaking practical research on CCWI in the Nakajima School District, an area around the Miya River in Ise City, which is located in the middle of Mie Prefecture (itself located in the middle of Honshu, the main island of Japan). Figure 3 shows the location of this study area, which has a population of 5,550 in 2,640 households (11/30/2017). The area is located in the lower region of the Miya River, whose headwaters are in the rainiest area of Japan. The Miya River is 91 km long with a basin area of 920 km^2, and 140,000 people live near the river, according to research by the Mie Office of Rivers and National Highways. This river flows toward the north along the western boundary of the study area. The average annual rainfall in the area is 1837.6 mm (average for 1981–2010 at the Obata observation station which is located near this study area), and this amount is less than the average of 3147.5 mm at Miyagawa Town's observation station, which is located in the upper area of Miya River, in the same period and a little more than the average annual rainfall in Japan during the same period, as recorded in the weather statistics of the Japan Meteorological Agency (JMA). However, precipitation across a wide area of the Miya River Basin, including the area near the headwaters, reaches the study area, so floods have happened several times in the past. Therefore, massive river construction projects to protect low-lying areas, such as those near the banks, have been undertaken through the ages. This district also has a little hill, Mt. Sango

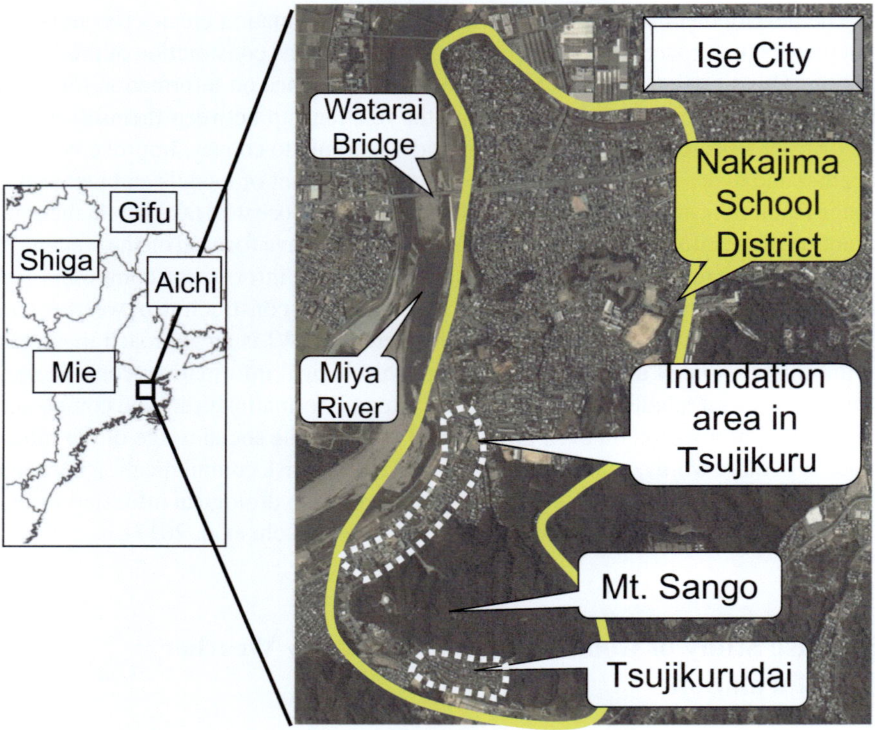

Fig. 3 Collaborative community weather information research area. *Created by the author based on GSI tiles (Seamless photo maps) (Geospatial Information Authority of Japan), 2018

(142 m), in the southern area, and the area around it is designated as a sediment-hazard area.

In the Nakajima School District, the residents themselves operate the three CCWI systems (December 2017) shown in Fig. 4 and prepare their local disaster risk strategies. These CCWI systems were created through collaboration with the local government, the weather office, the river manager, the community council, residents, and researchers. Through collaborative workshops, we collected local disaster knowledge from the residents and checked the relationship between this local disaster information and professional weather and hydrological information. (For details of the collaborative workshops, refer to Takenouchi et al. (2017)). As a result of this collaboration, the present CCWI systems were created. In the following, we give a brief explanation of each of the CCWI systems.

(i) Inundation disaster warning for preparation in Tsujikuru

When the water level in the Miya River beside the district rises, a floodgate is closed. Tsujikuru area has a risk of inundation because there is no water discharge with this closure. In the past, this area experienced inundation damage several times. This

Fig. 4 Collaborative community information in the Nakajima School District

i) inundation disaster warning for preparation in Tsujikuru*

*picture reference: document of the Mie Office of Rivers and National Highways.

ii) Watarai bridge information of exceeding purple Line

iii) sediment disaster warning on Mt. Sango behind Tsujikurudai

information system uses the amount of rainfall in the upper area when inundation happened in the past and the water level when the floodgate is closed.

(ii) Watarai bridge information of exceeding purple Line

On the pier of Watarai Bridge in the district, several water levels are painted in vivid colors. The opinion of residents that “when the water level exceeds the purple line, residents become conscious of the danger of a disaster” is reflected in this information. The level of this purple line is equal to the evacuation judgment level, so when the forecast predicts a water level requiring evacuation, this information is passed to residents near Miya River as a “level above the purple line.”

(iii) Sediment disaster warning on Mt. Sango behind Tsujikurudai.

Tsujikurudai housing complex is located at the base of Mt. Sango. Residents of the Tsujikurudai area have experienced dirty flow from Mt. Sango in the past as a result of heavy rainfall and worry about damage from sediment. This information functions to share the sediment disaster risk at this area. Sediment disaster risk depends on local conditions, so this system uses 5-km-mesh information that is produced by a highly technical forecast model dedicated to sediment disaster prediction.

We have observed the operation of the Nakajima CCWI system by the residents in a real case. Since the beginning of the utilization of CCWI, the members of the council, including representatives of local residents, have checked whether the meteorological situation meets the conditions of their CCWI system and, in such cases, the residents share the announcement of the CCWI by e-mail or telephone and communicate the risk in their community. The CCWI system was activated twice before but, fortunately, this district was not damaged in either case. The case to be discussed was the following. On September 9, 2015, the 18th typhoon of the season approached this region, but the Nakajima District was hardly affected. However, Kusube District, which is located at a distance of only about 5 km away, was hugely damaged by the local heavy rainfall. We verified that the conditions in Kusube District could have been applied in Nakajima District and checked the effects of the CCWI precautions. We confirmed that the CCWI could work effectively for the residents to prepare for the disaster. The community responses in this case are described briefly below.

At noon on September 8, information about the meteorological situation for the next few days was issued by the local weather office. The members in the councils discussed their preparedness based on this information. The information indicated that the situation was not severe: rainfall of 250 mm/day somewhere in the southern area of the prefecture and a pressure of 985 hPa with the arrival of the typhoon in this region. During the 21st typhoon of the 2004 season, which caused heavy inundation in this district, the amount of rainfall was about 700 mm/day. As a result of the discussion, they decided on continuous monitoring of the weather situation until the typhoon arrived the next day.

The next day, there had been strong rainfall around Ise since the early morning. At 07:40, the danger of a real-time landslide risk in this district reached the “Caution

Level." After receiving this information, the members in the council discussed and decided to share the CCWI on sediment disaster as a warning to prepare for evacuation. The members shared the situation to the residents by e-mail or telephone, and the community made their preparation to respond to the disaster. However, after 8 o'clock, the rainfall became weak, so the evacuation instruction from local government was not issued. In the damaged area of Kusube District, an evacuation advisory was issued at 08:17 and the evacuation order was given at 08:41.

We verified that if the rainfall situation that occurred in Kusube District had also happened in the Nakajima School District, an evacuation advisory would have been issued at 09:07 with a subsequent evacuation order at 09:31. So, the CCWI system could have prepared this community prior to the more severe disaster risk situation.

As shown in the above case, the residents could use the important weather and hydrological information for themselves by themselves and deal with the local disaster risk. There was, therefore, effective and active use of weather and hydrological information that was formed in the community through the CCWI system (Takenouchi et al. 2017).

6 Summary and Recommendations

In this chapter, we have considered a future vision of weather and hydrological information involving risk communication among various stakeholders. When the information is primitive, it is easy for residents to have a relationship with the information, and residents are expected to have a personal awareness of disaster risk. However, when the information is highly developed, it is difficult for residents to participate in its construction and their awareness declines as a result of depending too much on the provision of information by others. As the Sendai Framework for Disaster Risk Reduction indicated, residents' participation is significant to reflect traditional knowledge and local characteristics in weather and hydrological information. In addition to understanding the significance of this, we need to be careful about the possibility of differences in objectives and methods of response to local disaster risks in community situations as shown in Sects. 1 and 2.

To deal with such situations, CCWI was proposed as a social system in which residents have a relationship with highly technical weather and hydrological information. As an example, the case study in Nakajima School District was introduced, which showed the possibility of using highly developed and highly sophisticated information in relation to local disaster risk through the use of real CCWI systems. However, the method introduced in this practical research also has some issues. Each community has its own local knowledge and disaster risks. Considering the characteristics of each community, whether specific CCWI systems for every community can be socially created is a big issue. A lot of effort is required to ensure that relevant stakeholders collectively consider CCWIs for their community. So, it is necessary to consider methods to create CCWI systems efficiently by agreement in each community. In Japan, the Community Disaster Management Plan was regulated and has been executed since April 2014. This plan is a rare system that is not found in other countries. This system defines residents' efforts in their community for disaster prevention by law and is very familiar to CCWI. If a social structure that promotes risk communication based on weather and hydrological information is grown, we will be able to broaden areas where CCWI systems work in communities. We also should not forget the existence of science and technology, which can develop high-resolution models and make the use of them in the community easier.

In the future, occurrences of extreme meteorological phenomena and unexpected disasters are expected to become more frequent as a result of global climate change. This will enhance the importance of the basic maxim "help thyself" in preparing for future disasters. To support residents to take proper action smoothly, we need to construct a social system for disaster prevention which advances disaster risk communication.

References

Almoradie A, Cortes VJ, Jonoski A (2015) Web-based stakeholder collaboration in flood risk management. J Flood Risk Manag 8:19–38

Bateson G (1972) Steps to an ecology of mind. Jason Aronson Inc., London

Catherine CA, Carlos Primo CD, Guillermo QT III (2012) Community-based monitoring for flood early warning system: an example in central Bicol River basin, Philippines. Disaster Prev Manag: Int J 21(1):85–96

Chen SC, Wu CY (2014) Debris flow disaster prevention and mitigation of non-structural strategies in Taiwan. J Mt Sci 11(2):308–322

Gautam DK, Phaiju AG (2013) Community based approach to flood early warning in West Rapti River Basin of Nepal. J Integr Disaster Risk Manag 3(1):155–169

GDPFS (2015) WMO technical progress report on the global data-processing and forecasting system and numerical weather prediction research. http://www.wmo.int/pages/prog/www/DPFS/ProgressReports/2015/GDPFS-NWP-2015.html. Accessed 25 Mar 2018

Government of Japan (2016) The 5th science and technology basic plan. http://www8.cao.go.jp/cstp/english/basic/5thbasicplan.pdf. Accessed 25 Mar 2018

Guha-Sapir D (2018) EM-DAT: the emergency events database—Université catholique de Louvain (UCL)—CRED, Brussels, Belgium. www.emdat.be. Accessed 25 Mar 2018

IFRC (International Federation of Red Cross and Red Crescent Societies) (2015) World disasters report. http://ifrc-media.org/interactive/wp-content/uploads/2015/09/1293600-World-Disasters-Report-2015_en.pdf. Accessed 25 Mar 2018

IPCC (Intergovernmental Panel on Climate Change) (2012) Managing the Risks of Extreme Events and Disasters to Advance Climate Change Adaptation. A Special Report of Working Groups I and II of the Intergovernmental Panel on Climate Change In: Field CB, Barros V, Stocker TF, Qin D, Dokken DJ, Ebi KL, Mastrandrea MD, Mach KJ, Plattner G-K, Allen SK, Tignor M, Midgley PM (eds) Cambridge University Press, Cambridge, UK, and New York, NY, USA, 582 p

Japan Meteorological Agency (2011) Report of the result of questionnaire survey "Survey on utility situation of weather information for disaster prevention", 05/2011, Japan Meteorological Agency. http://www.jma.go.jp/jma/press/1105/31a/manzokudo_data.pdf (in Japanese). Accessed 25 Mar 2018

Katada T (2006) Study on the way of social co-learning based on disaster investigations and the results, working papers of FY2005 "Priority research issue (Research grants)" (in Japanese)

Katada T, Kodama M, Kuwasawa N, Koshimura S (2005) Issues of resident's consciousness and evacuation from the tsunami -from questionnaire survey in Kesennuma city, Miyagi Pref. After the earthquake of Miyagi-oki, 2003. In: Proceedings of Japan society of civil engineers, no 789/II-71: 93–104 (in Japanese)

Macherera M, Chimbari MJ (2016) A review of studies on community based early warning systems. Jàmbá: J Disaster Risk Stud 8(1):1–11

Maskrey SA, Mount NJ, Thorne CR, Dryden I (2016) Participatory modelling for stakeholder involvement in the development of flood risk management intervention options. Environ Model Softw 82:275–294

Sasada K, Yi-Tzu Lin YT, Sato N (2015) Development of non-structural measures toward dangerous area for debris flow disaster in the mountainous area of Taiwan and implications for Japan. J Jpn Soc Nat Disaster Sci 34(3):189–211 (in Japanese)

Smith PJ, Brown S, Dugar S (2017) Community-based early warning systems for flood risk mitigation in Nepal. Nat Hazards Earth Sci 17(3):423–437

Takemi T, Okada Y, Ito R, Ishikawa H, Nakakita E (2016) Assessing the impacts of global warming on meteorological hazards and risks in Japan: Philosophy and achievements of the SOUSEI program. Hydrol Res Lett 10(4):119–125

Takenouchi K, Kawata Y, Nakanishi C, Yamori K (2014) Collaborative community weather information-sharing disaster risk between producer and user-. J Disaster Inf 12:100–113 (in Japanese)

Takenouchi K, Nakanishi C, Yamori K, Sawada M, Takeuchi K, Fujiwara H (2017) A trial of collaborative construction of local weather information at Nakajima school district in Ise. J Integr Disaster Risk Manag 7(2):1–24

United Nations (2015) Sendai framework for disaster risk reduction 2015–2030. http://www.unisdr.org/we/coordinate/sendai-framework. Accessed 25 Mar 2018

Yamori (2013) Risk communication in huge disasters-a new style of disaster information. Minerva Publishing (in Japanese)

Disaster Education Based on Legitimate Peripheral Participation Theory

A New Model of Disaster Science Communication

Takuya Iwahori

Abstract It is often suggested that disaster education should not be a one-way knowledge transfer from disaster experts to non-experts, but a bilateral interaction between the two sides. In this study, the authors propose a new framework for disaster education based on legitimate peripheral participation (LPP) theory, in which disaster experts and non-experts interact very closely to remove the barriers between the two sides. Specifically, the present study introduces two practical research methods for disaster education. The first is an attempt to convert a seismological observatory into a disaster science museum with collaboration between seismologists (experts) and volunteer staff (non-experts). The second is an endeavor to involve elementary school children in cutting-edge seismological research by installing a miniature seismometer at their school. As a result, volunteer staff at the museum formed a new identity in joint practice as semi-experts who mediate between experts and non-experts. Although the schoolchildren did not reach a definite stage as successfully as did the volunteer workers in the museum project, through this research, both the seismologists and the children realized what they had not shared, and this marked a starting point for further risk communication. The results are discussed from the perspective of LPP theory.

Keywords Disaster education · Legitimate peripheral participation · Community of practice · Science communication · Seismology

1 Introduction

Recent natural disaster research suggests the growing importance of the role of non-experts in disaster prevention activities. This paper analyzes the relationship between experts and non-experts during disaster prevention-related activities, with a particular focus on disaster education. The first purpose of this research is to postulate

This chapter is a revised version of Iwahori et al. (2017).

T. Iwahori (✉)
Disaster Prevention Research Institute, Kyoto University, Kyoto, Japan
e-mail: iwahori@drs.dpri.kyoto-u.ac.jp

K. Yamori (ed.), *Disaster Risk Communication*, Integrated Disaster Risk Management,
https://doi.org/10.1007/978-981-13-2318-8_6

that, in these situations, knowledge transfer is not central to learning, but that the relationship formed between experts and non-experts constitutes learning in itself that can be interpreted as disaster education. Consequently, this paper evaluates the effectiveness of learning that occurs among experts and non-experts in environments where purposeful interaction on the topic of disaster preparedness is introduced. The second purpose of this research is to evaluate the effectiveness of this type of disaster education where knowledge transfer is not central.

After a number of recent major catastrophic natural disasters in Japan, including the Great Hansin Awaji Earthquake in 1995 and the Great East Japan Earthquake in 2011, we often hear statements such as "modern high-tech structures in Japan were thought to be undamageable by earthquakes, but the idea that an elevated highway cannot collapse is a 'safety myth'," and "seismologists try to make excuses by saying the size of the tsunami was totally 'unexpected'" (as examples, see *The Japan Times* 2004; Okada et al. 2011; Nogami and Yoshida 2014). The terms "unexpectedness" and "safety myth" have one important aspect in common, that is, an obvious lack of mutual communication and understanding between disaster experts and non-experts. Such phrases do not simply point to certain failures/defects in disaster science/technology in itself, but to the fact that experts have not successfully communicated with ordinary people about the limits and outcomes of their research.

For example, Kumamoto (2011) indicated that, even before 2011, some seismologists warned of the possibility, albeit very slight, of an unprecedentedly large tsunami in the Tohoku region based on research on the Jogan Earthquake in 869 CE. However, this warning was not taken very seriously in the academic or governmental sector. As a result, the warning was not widely communicated to non-experts. This poor risk communication generated a feeling of "unexpectedness" and "unknownness" in the aftermath of the Great East Japan Earthquake; however, in reality, the tsunami was not unexpectedly or unprecedentedly massive.

2 Previous Studies

Before the Great Hanshin Awaji Earthquake of 1995, disaster prevention in Japan had mostly been based on hard engineering solutions. However, following this disaster, the importance of people-based approaches and risk communication became widely recognized in Japan, and many actions were taken in that area. Disaster education is one of the areas that came to prominence at the time as a people-based approach.

Disaster education is closely related to disaster risk communication. Risk communication theory suggests a gap between experts and non-experts should be carefully dealt with when experts try to transfer knowledge to non-experts. It is true that knowledge transfer is needed as a basic step of disaster education, but the style of knowledge transfer should be taken into careful consideration (Yamori et al. 2007; Shiroshita and Yamori 2011), because knowledge transfer sometimes piles up an impenetrable barrier between experts and non-experts.

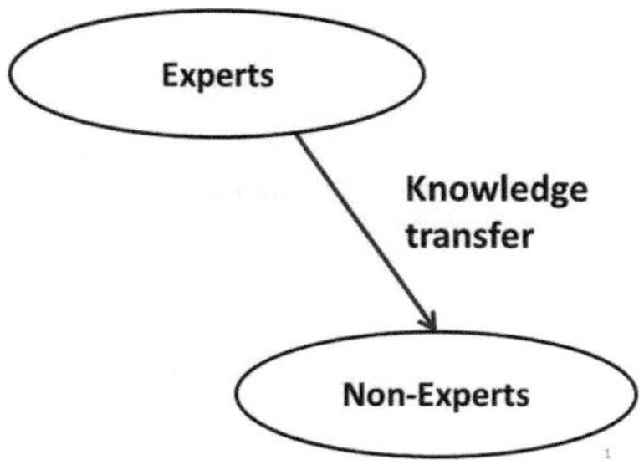

Fig. 1 Deficiency model

One of the most sharply contrasted features in the argument of disaster education is top-down/bottom-up, expert-centered/citizen (or non-expert)-centered, or directive/participatory approach. However, communication gap between experts and non-experts is not filled up easily, even if we take more participatory approaches which look more suitable for this purpose. This is mainly because both types of approaches are within the realm of knowledge transfer paradigm. We must step out from such a conventional way of thinking, but focusing more on restructuring identities of those who transfer knowledge (experts or instructors) and those who receive knowledges (non-experts or learners) in knowledge building.

The problem pointed out by Yamori et al. (2007) and Shiroshita and Yamori (2011) in disaster education is known as the "deficiency model" in science communication theory. The "deficiency model" assumes "experts to be on the side of scientific knowledge and the general public to be on the side of no scientific knowledge, that is, on the side that is deficient in scientific knowledge. Information is assumed to be transmitted from experts with scientific knowledge to citizens with a lack of scientific knowledge" (Fujigaki and Hirono 2007). Figure 1 illustrates the deficiency model.

3 Use of Legitimate Peripheral Participation (LPP) Theory

This paper uses LPP theory to explain how learning occurs among participants who do not interact with the specific purpose of knowledge transfer, but rather with the purpose of learning sharing and information exchange. In LPP theory, learning is formulated as "legitimate peripheral participation in a community of practice." From this definition, we introduce technical terms and the relationship between participants in the learning process. As shown in Fig. 2, in the "community of practice," full participants and new participants, both of whom are categorized as peripheral participants, maintain mutual negotiation on the basis of a central practice.

LPP theory interprets learning as identity building of each participant rather than knowledge transfer between the two. Building an identity consists of negotiating the meaning of our membership experiences in the community of practice.

Fig. 2 Model of learning based on legitimate peripheral participation (LPP) theory

3.1 LPP

(1) Theory

Lave and Wenger (1991) first proposed LPP theory, which assumes that the essence of learning is LPP in a community of practice.

An explanation of the technical terms in LPP theory is as follows: "Community of practice" is a group of people involved in a specific practice; "legitimate peripheral participation" refers to an "authentic practice" of a community of practice in a plural and diverse way that is possible to perform, and the changing of the relationship itself with the passing of time. Here, the word "legitimate" means "authentic," "peripheral" means "plural" and "diverse," and "participation" literally means "participation" that is concretely performed.

These definitions are in accordance with Yamori's (2012) interpretation. In the present study, we revised the theory according to the definitions in the following paragraphs.

(2) Conditions of LPP

In this section, we consider the three conditions—legitimate, peripheral, and participation—for realizing LPP. These three conditions can be integrated into the following two parts: "legitimate and peripheral" and "participation."

The condition "legitimate and peripheral" means that there is a community of practice whose "authentic" practice involves a "diverse" form of participation by existing members. To put it another way, "legitimate and peripheral" means that a community of practice is "accessible." Shiroshita and Yamori (2011) is correct when he says that "'legitimacy' is the problem of whether or not there can be participation in a practice. 'Peripherality' is the problem of whether or not access is allowed to a valuable activity in the practice concerned." Legitimacy or what is regarded as "legitimate" is beyond the scope of our current discussion.

The other condition, "participation," means that each specific community of practice has active participants. It is easy to understand this condition because "participation" literally means participation. Here, we notice that the condition of "participation" is more concrete than the condition of "accessible," which is very abstract. Shiroshita goes on to say that "and if it is guaranteed and participation is done in a community of practice, learning is generated in itself." What the passage makes clear at once is that two conditions, "participation" and "accessible," comprise learning in LPP.

It follows that, to realize LPP, it is necessary that a community of practice be "accessible" and have active participants.

(3) Concept of "peripheral"

The concept "peripheral" means "diverse" or "plural." It is key to understanding LPP theory. It should be emphasized that it is a mistake to think of a dichotomy between "central" and "peripheral" in understanding the theory in terms of the plurality and diversity of "peripheral" in contrast with the unity symbolized by the concept of "central." The idea of "peripheral" represents diversity and plurality by subcategorizing various forms of "peripheral participation." Since all forms of existing participation are categorized as LPP according to the theory, the concept of "peripheral" represents a unity of the idea, in contrast to the plurality and diversity represented by its meaning.

We can represent the constitution of learning by setting a "central practice" in a community of practice. The term "central practice" can be defined as "a certain practice that is symbolic to a specific community of practice." We can set a central practice selectively from various practices that are observed in the community of practice, according to the focus of analysis. By doing so, we can analyze learning arising from LPP by a community of practice, whose central practice is not the transfer of knowledge and techniques that is generally regarded as learning.

After we select a central practice of a specific community of practice, any (legitimate) participation in the community of practice can be categorized as peripheral participation. Since the term "peripheral" means diverse, any possible (legitimate) participation in the community of practice can be interpreted as diverse participation.

Here, we have two distinct types of peripheral participants, namely, full participants and new participants. These two subcategories of peripheral participants are defined in a correlative way. According to the central practice specified, participants who have more experience become full participants. Conversely, people who have no experience, or at least less experience than full participants, become new participants. By subcategorizing these two kinds of participants, we can consider peripheral participation in itself, not in a way based on the commonly used binominal opposition of "central (participants)" and "peripheral (participants)."

All existing participants can be categorized as full or new participants. Future participants, who do not exist in the context of central practice at the time of the distinction between full participants and new participants, can be excluded from the subcategories. However, they may also participate in it in the future as new participants. Therefore, future participants may also be categorized as peripheral participants.

Various participants can be integrated into a united form of "peripheral participants." With such a categorization, a consistent view can be seen in analyzing a participation process that changes with the passage of time.

(4) Discourse of learning

The main point regarding the discourse of learning through LPP is continuous mutual negotiation between full and new participants mediated by a central practice, as shown in Fig. 2.

According to Lave and Wenger (1991), continuous negotiation between the two participants causes three different kinds of change as a result of learning through LPP: new participants move toward becoming full participants, while full participants come to accept the "constructively naïve perspectives or questions" that new participants bring in from outside the community of practice as "inexperience [that is] is an asset to be exploited." Acting as an intermediary between full and new participants, the community of practice itself changes.

3.2 *Identity-Building Evaluation of Learning*

We are now ready to consider how to evaluate the results of learning engendered by LPP. We interpret the three results mentioned in the previous subsection as identity building by participants.

Lave and Wenger (1991) believe that participation in a community brings about self-awareness of becoming part of that community, as can be seen in the following quotation:

> Moving toward full participation in practice involves not just a greater commitment of time, intensified effort, more and broader responsibilities within the community, and more difficult and risky tasks, but, more significantly, an increasing sense of identity as a master practitioner.

This idea marks a clear contrast with the traditional idea that regards the results of learning to be the acquisition of knowledge and/or techniques.

Yamori (2012) has the following to say on the matter of identity building:

> We shall concentrate on the concept of "identity" stated by Lave and others. They use the word identity not in the narrow sense of personality or ego identity of individuals, but in the sense of the role or position that participants take in a community of practice .... The important point of learning is to obtain membership in a community in which participants act cooperatively.

Relevant to this point is Takagi's (1999) remark:

> In LPP theory, "to participate" corresponds to a situation in which the participant is in charge of a role of a certain practice accepted as legitimate in the community of practice. Therefore, we can interpret learning as a change in the way of participating, that is, a change in the role the participant takes in the community of practice.

From these remarks, one general point becomes very clear: The result of learning through LPP appears as a change in the role of the participant in a community of practice. We can understand the change in the role as the participant's change in identity. At the same time, the changes in role and identity are balanced by the change in the community of practice.

4 Case Study

Theoretical assumptions are tested through case studies. Here, we introduce one scientific research project, the Manten Seismic Observation Project, and two field studies, the Abuyama Open Laboratory and a collaborative study program between an elementary school and the Manten Seismic Observation Project. Both case studies are characterized by the participation of non-experts in the practice of experts.

4.1 Abuyama Open Laboratory

The Abuyama Open Laboratory is an attempt to convert a seismological observatory into a disaster science museum with collaboration between seismologists (experts) and volunteer staff (non-experts).

The Abuyama Open Museum Project started in December 2010. The core members of the project are Yamori and Iio, who are Professors at the Abuyama Observatory, and a Planning Director at the Disaster Reduction and Human Renovation Institution. The Abuyama Open Museum Project aims to transform the Abuyama Observatory from a center of data collection for seismology into a museum of seismology open to the general public.

During the startup period, the policy of the project was stated as follows: "When we think about the future of this observatory (building institute), it is the best decision to transform the observatory into a museum that mainly focuses on the history of seismology. With such a transformation, we believe that the observatory could be all the more valuable, in that it would be open to the general public, be a place to communicate with locals, and effect various types of learning, resulting in greater satisfaction by introducing numerous people to studies on seismology and other subjects related to the Abuyama Observatory."

Abuyama Open Laboratory is a series of events that were carried out based on such a policy. Abuyama Open Laboratory consists of two kinds of public events. One is monthly open houses, and the other is special seasonal open houses. The open house schedule is announced publicly, and citizens who wish to visit Abuyama Open Laboratory apply to participate in the event according to the schedule. In addition to a guided tour, several seminars or courses are arranged as part of the basic open house program according to the theme set each time such an event is held.

Abuyama Open Laboratory was held for the first time on April 3, 2011, not long after the Great East Japan Earthquake on March 11, 2011. The catastrophic disaster caused a certain unrest in Japanese society. This mood influenced the core members, making them hesitate about whether to hold the first special open house. Although they eventually decided to carry out the event on schedule, the scale of the event was reduced.

Abuyama Open Laboratory's second special open house was held on July 31, 2011. On this day, programs were arranged for children in the midst of their summer vacation and their parents. Since the programs were favorably received and successfully completed, core members began to consider Abuyama Open Laboratory more seriously.

The third special open house of Abuyama Open Laboratory was held on November 5, 2011. On this day, dialogical courses between core members and participants were arranged. The main themes of the dialogue set in the programs were the possibility of development of Abuyama Open Laboratory in the future. As a result of the dialogue, core members shared the common purpose of introducing volunteer supporters.

A volunteer supporter training course was arranged for June 2–3, 2012 as a special open house. On these two days, participants joined training programs necessary for leading a guided tour of the Abuyama Open Laboratory. For example, they studied old seismometers, the history of the observatory and Mt. Abuyama, and the accepted protocol for receiving guests. Participants who finished the training program were authorized as volunteer supporters.

The second volunteer supporter training course was arranged for October 20–21, 2012 as a special open house. On these two days, volunteer supporters who had continued training since the previous volunteer supporter training course practiced giving guided tours to visitors for the first time.

Hereafter, volunteer supporters began to lead guided tours of Abuyama Open Laboratory. In the 2013 academic year, volunteer supporters led 10 tours.

The third volunteer supporter training course was arranged for June 22–23, 2013 as a special open house. New volunteer supporters were recruited and participated in the training noted above.

Particularly after the year of 2013, volunteer supporter has expanded the scope of their activities greatly. Supporter's central role has been a guide and lecturer for visitors in a guided tour in the observatory. The times of guided tours supporters planned, organized, and engaged has been dramatically hiked: 0 in the year of 2011, 16 in 2012, 53 in 2013, 40 in 2014, 65 in 2015, 83 in 2016, and 92 in 2017. The total number of visitors has been changed: 801 in the year of 2011, 802 in 2012, 1712 in 2013, 1144 in 2014, 1661 in 2015, 1960 in 2016, and 2332 in 2017. This fast development clearly and impressively shows that volunteer supporter who are originally normal citizens, i.e., non-experts, now take an indispensable rule in science communication in the observatory, bridging science and society, that is, experts and non-experts.

4.2 Manten Seismic Observation Project

The Manten Seismic Observation Project was an attempt to densely observe local inland earthquakes using hundreds of small seismometers. Iio's research group developed the latest version of the Manten seismometer in 2008. The Manten seismometer allows offline seismic observation. It consists of a velocity-type seismograph, a data logger, and a battery. The Manten seismometer is characterized by its power efficiency, which enables data to be recorded for up to about nine months. Another notable feature of the Manten seismometer is that many units of Manten seismometers are installed at the same time near each other, and so these are chosen for intensive observation of microearthquakes to clarify earthquake mechanisms.

The Manten seismometer is installed at the same time at many points in nearby areas, which are chosen to allow intensive observation of microearthquakes in order to clarify the mechanisms underlying earthquakes. For example, 82 observation points were set in the northern part of the Kinki area and 50 in western Tottori Prefecture; there are also observation points in eastern Shimane Prefecture, western Nagano Prefecture, the Nobi area, the Kyushu area, and New Zealand. As of July 2014, a total of 250 observation points using the Manten seismometer had been set to observe microearthquakes.

In Japanese, the word *manten* has two meanings that relate to the Manten Earthquake Recording Project. One meaning is "tens of thousands of points," which represents the ideal image of seismic observation. The other meaning is "perfect score," which means that the seismic observation project is (one of) the best ways to observe earthquakes.

It should be noted that volunteer supporters have started to participate the Manten Earthquake Project itself as semi-scientific staff, along with their activities as science communicators within the observatory. For one, supporters engaged installment procedure of new seismometers, searching for a proper site, fixing it properly, and doing necessary maintenance works. An impressive accomplishment was their activities in the aftermath of the 2018 Northern Osaka Earthquake, hitting Osaka Metropolitan area, registered M6.1. It was deadly needed to start to an emergent observations of a series of aftershocks in order to be prepared well for possible major aftershocks. Supporters were quickly involved in an emergent observation team organized by university researches and contributed greatly to a quick setup of network of seismometers. This immediate and productive action was possible thanks to trust relationship between experts (seismologist) and supports (citizens), steadily constructed through previous collaborative works in a diverse channels.

For another, volunteer supporters are also engaged in data cleaning and data mining works under the supervision of expert seismologists. A bulk of data gathered in the Manten Earthquake Project needs data cleaning process of eliminating noise data caused by car traffics, for example. The process can be done mainly by automatic cleaning procedure, but it needs final checkup by humans. Now, supporters are well trained and qualified to do this important works in a whole earthquake science process.

Active involvement of volunteer supports in the Manten Earthquake Project outlined above clearly shows supporters who are originally normal citizens, i.e., non-experts, now take an integral part of science itself, not just an inter-mediator between science and society, that is, experts and non-experts.

4.3 Collaborative Study Program Between an Elementary School and the Manten Seismic Observation Project

A collaborative study program between an elementary school and the Manten Seismic Observation Project attempted to involve elementary school children in cutting-edge seismological research by placing a miniature seismometer in a school.

(1) Shimoyama Elementary School

Shimoyama Elementary School is located in the mountainous area of central Kyoto Prefecture. It has about 80 pupils (note: Japanese elementary schools have six grades).

A Manten seismometer was set up in Shimoyama Elementary School in 2009. Two motives combined to lead Yamori to install a seismometer at the school site. First, it would contribute to progress of the latest seismological research of the Manten Seismic Observation Project by adding another observation point. Second, it would enrich disaster education within the school and aid the shared practice of seismometer maintenance between experts and pupils.

After approval by the School Board, the plan to set up a Manten seismometer in Shimoyama Elementary School was realized on December 8, 2009. The actual work of installing the Manten seismometer was conducted by experts and pupils.

There is a network of Manten seismometers in the northern Kinki area. The Manten seismometer set up in Shimoyama Elementary School was a new observation point in the network. After the installation of the Manten seismometer, we continuously carried out disaster education. The class schedule was set as four times a year, in accordance with the timing of maintenance of the Manten seismometer. The classes consisted of seismometer maintenance by the pupils and the authors, and classroom lessons with themes related to the Manten seismometer and the Manten Earthquake Recording Project.

(2) Neu Elementary School

Neu Elementary School is located in a mountainous area of western Tottori Prefecture. It holds around 100 pupils. The Manten seismometer was set up in Neu Elementary School in 2010.

In 2000, the Western Tottori Earthquake occurred, striking Hino Town, where Neu Elementary School is located. It was around this area where damage from the earthquake was most extensive.

In 2010, the Tottori Prefectural Government planned a series of projects in accordance with the 10-year anniversary of the Western Tottori Earthquake. One of these projects aimed to promote disaster education at Neu Elementary School. Neu Elementary School was selected because, first, Yamori was looking for a school in which he could carry out a disaster education project. Second, there is a connection between the Yamori and Tottori Prefectural Governments. Third, Neu Elementary School is located within the area of a network of Manten seismometers.

Iio planned to set a new seismometer at the site of Neu Elementary School because it is located in a 50 km^2 area that already contained 50 observation points. However, according to firsthand observations, if a new Manten seismometer had been set up in that location, noise from a nearby railway, a national road, and a river would have been excessive. Therefore, a site two kilometers from the school was chosen as the installation point. On May 17, 2010, Iio and students from Neu Elementary School installed the Manten seismometer.

Neu Elementary School has disaster education study programs other than the one in collaboration with the Manten Seismic Observation Project. One of these is listening to experiences of the Western Tottori Earthquake from family members, locals, and volunteer staff who have worked on the revitalization of the region. Other examples include a presentation in a public disaster forum and an annual school recital about what they have learned through investigative learning. The Tottori Prefecture Board of Education evaluated these various types of disaster education and gave an official commendation to the school in 2011.

In both of these elementary schools, we continued the disaster education program after the installation of the Manten seismometer. The disaster education program consists of three or four classes a year. The purpose and practical significance of the program are as follows. First, in the study program, elementary school pupils handle the maintenance of the Manten seismometer, even though they are generally thought to be practicing disaster prevention activities. Second, it is becoming difficult for seismology experts to periodically look after the seismometers as they spread out from the center of research.

As of July 31, 2014, we had held 17 classes at Shimoyama Elementary School and 15 at Neu Elementary School. In this paper, we focus our attention on seven classes held in each elementary school in the period of 2011 and 2012. We briefly summarize these classes in the paragraphs that follow. Sentences in which the name of the elementary school is not clear represent activities that were commonly performed in both schools.

Classes were held four times a year and attended by sixth-grade students. The class schedules were arranged in accordance with seasonal maintenance of the Manten seismometers. Each class consisted of seismometer maintenance and classroom lessons. Maintenance of the Manten seismometers set in the elementary schools was performed by the pupils and the authors. Classroom lessons were held under themes related to the Manten seismometer and the Manten Earthquake Recording Project. The practical purpose of the classes was to have pupils get a proper feel for the role of an expert.

The work of seismic observation was divided between the pupils and the experts as follows. The pupils were in charge of seismometer maintenance and the experts offered support. The experts, as usual, did the rest of work, which included reading the data of collected CompactFlash (CF) memory and evaluating underground structures by analyzing the data.

Now, let us look closely at the work of seismometer maintenance. Seismometer maintenance is divided into three procedures: "ending procedure for seismic observation," "exchanging batteries and CF card memories," and "restarting seismic observation." Every year, in both elementary schools, about 10 sixth-grade pupils participated in seismometer maintenance. Accordingly, the pupils were divided into three groups, with each group in charge of one of the three procedures mentioned above.

Examples of the study program are provided in the following paragraphs.

In 2011, the constitution of the study program was irregular in that it specially focused on the Great East Japan Earthquake, which occurred just before the start of the elementary school year and shocked Japanese society. In the first class of the year in May, the first author taught sixth-grade pupils about the Great East Japan Earthquake and how it related to the Manten Seismic Observation Project. In the second class in September, pupils made handicrafts of the seismometer using PET bottles. Through the handicrafts, they learned the basic mechanisms of the seismometer. In the third class in November, classes were held under the theme of the role of experts who lead the Manten Earthquake Recording Project, including how these roles relate to the pupil's role of seismometer maintenance. In the fourth class in February, a group of pupils and teachers from Shimoyama Elementary School visited Abuyama Observatory. Through this visit, they deepened the knowledge that they had acquired in previous classes. Neu Elementary School is far from Abuyama Observatory, so instead of visiting, these pupils performed group work involving the reading of waveform data. In this group work, waveform data collected from the Manten seismometer were printed, and pupils then distinguished the waveform data of earthquakes from those of noise such as vibrations of passing trains or strong rainfall.

In 2012, the constitution of the study program was basically the same as what it had been in the previous year. The only major change from the previous year is that, in 2012, the fourth class was an online class linking Neu Elementary School, Shimoyama Elementary School, and Abuyama Observatory through a teleconference system. We planned it specifically for Neu Elementary School because they could not visit the faraway Abuyama Observatory. Through this class, each elementary school and Abuyama Observatory shared information about their activities associated with the Manten Seismic Observation Project and the unique culture of each place, which allowed both the pupils and the staff of Abuyama Observatory to deepen their understanding of the study program.

5 Discussion

In the community of practice, Abuyama Open Laboratory experts (full participants) and volunteer supporters (new participants) maintained an ongoing mutual negotiation around guided tours of the observatory. This resulted in the volunteer supporters building up a new identity and letting the experts rediscover the fundamentals of their roles.

In the community of practice of the "study program in elementary schools in collaboration with the Manten Seismic Observation Project," an LPP analysis of the apparently passive attitude of the students identified a constructive meaning.

According to LPP theory, participating in a community of learning where knowledge transfer is not central to learning, an "ideal" central participant (Fig. 5) does not respond to an expert, as argued in the deficiency model of risk communication theory. In other words, the central participant is not negative toward learning. Participation in the "community of practice," where knowledge is shared rather than transmitted in one direction from experts to non-experts, becomes a catalyst for shared learning.

5.1 Change in Identity of Volunteer Supporters

A change in identity among volunteer supporters resulting from learning through LPP appeared in the following respect: the participants in Abuyama Open Laboratory took charge of the role of volunteer supporters in collaboration with the experts.

Now, let us look closely at the discourse of learning among volunteer supporters. To begin with, at first, as visitors, the participants attended a volunteer supporter training course until becoming qualified. Next, these volunteer supporters continued taking voluntary training accompanied by seismology experts in a guided tour of the observatory. The relationship between the volunteer supporters and the seismology experts corresponded to mutual negotiation (interaction) between new and full participants, as mentioned in chapter "Using Computer Simulation for Effective Tsunami Risk Communication". Finally, the volunteer supporters led guided tours of the observatory for visitors.

The volunteer supporters who were at first visitors made efforts to take on the role of "volunteer supporters" who are mainly in charge of leading guided tours for other visitors. The volunteer supporters, by performing this role while still somewhat new visitors, realized that the visitors' own learning gives meaning to the guided tour.

LPP theory can explain how the role of volunteer supporters in the Abuyama Open Laboratory came into existence through their participation. In LPP theory, the result of learning is evaluated as a participant's identity building; here, identity means the role or position in the community of practice. Correspondingly, it can be assumed that the visitors' participation in the community of practice at Abuyama Open Laboratory resulted in identity building as volunteer supporters by taking and performing this role.

It is helpful to describe the role of volunteer supporters in Abuyama Open Laboratory before moving on to analyze the learning process. Their main task was to lead a guided tour of Abuyama Observatory. The guided tour is a central practice of the community of practice of Abuyama Open Laboratory. Other related practices include event receptions, making teaching materials and event exhibits, cleaning, and participating in voluntary training for guided tours.

Some data show the presence and scale of the activities. All data are as of July 31, 2013. First, 15 volunteer supporters continuously participated in the Abuyama Open Laboratory after qualifying. These volunteer supporters included 12 males and three females. Most of them were between the age of 60 and 80 years, and were past retirement age. A total of 211 people participated as volunteer supporters.

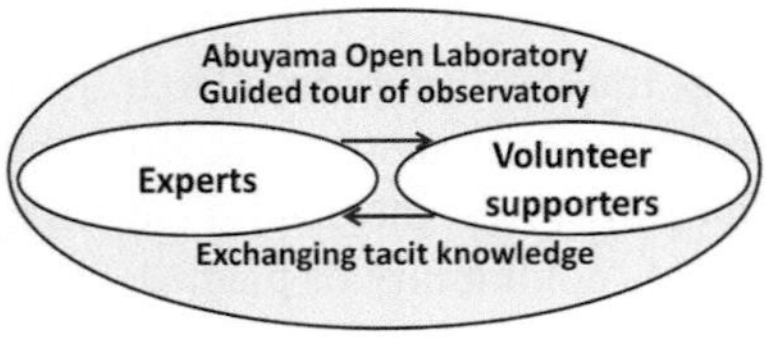

Fig. 3 Mutual negotiation between experts and volunteer supporters

An analysis of the mutual negotiation between the volunteers and the visitors (Fig. 3) shows that the volunteer supporters took their role as their identity, thus fulfilling two theoretical conditions of LPP, namely, that volunteer supporters exist and that the community of practice of Abuyama Open Laboratory is accessible to them. In demonstrating their expertise, the full participants have established their legitimacy. Here, we can also think plainly that the legitimacy of the full participants means they are generally regarded as experts in this society. To discuss legitimacy in its entirety is beyond the scope of this discussion, but we may say in this case study that both ways of thinking influenced the practice.

From a practical standpoint, in order for a guided tour to function effectively, a certain important precondition should be met; namely, those acting as volunteer supporters need to carry out guided tours that consist of too many visitors for experts only to handle. Another important precondition is the arrival of laboratory visitors for guided tours that are led by volunteer supporters. Furthermore, it should be noted that the mutually dependent relationship between volunteer supporters and visitors can be changed through the participation process, because it is also possible for visitors to deepen their learning through maintaining participation in the Abuyama Open Laboratory and becoming new volunteer supporters.

This argument can be expanded into an analysis of the mutual negotiation between experts and visitors (Fig. 4). When volunteer supporters take part in voluntary training for guided tours, they receive a transfer of seismology knowledge from experts. Considering this relationship, we can find another aspect of the role of volunteer supporters, that is, volunteer supporters accept the knowledge of experts and pass it on to visitors. To realize this role, the existence of experts is necessary.

Focusing on mutual negotiation, it is at first sight a simple transfer of knowledge or techniques. In fact, the volunteer supporters had to not only accept knowledge of seismology from experts, but also look for what they needed to know or do in guided tours. This process is considered to be the sharing and exchanging of tacit knowledge. The term "tacit knowledge" is used in the sense that it is a kind of knowledge or technique with which one can perform a certain practice, although

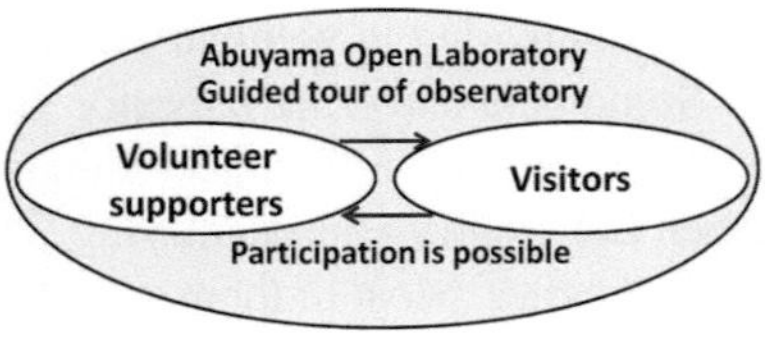

Fig. 4 Mutual negotiation between volunteer supporters and visitors

they are usually not conscious or have no clear perception of it (e.g., speaking or riding a bicycle).

For example, take the situation of voluntary training for guided tours. There were volunteer supporters, experts, and an old historical seismometer. In explaining the function of a pendulum in the seismometer, the experts used the term "natural period" (the period of one complete oscillation of a body or system) because they thought the meaning of the term was obvious. However, the volunteer supporters replied, "What is a natural period?" because they did not know this concept. In response to the question, the experts explained the natural period to the volunteer supporters. Later on, the volunteer supporters made a miniature pendulum, through which they explained the natural period to visitors in a guided tour.

They knew about the existence of a pendulum in the seismometer because they had participated in guided tours as visitors. But they did not know that the concept of the "natural period" was related to the function of the pendulum because they were not experts, and this idea is only common among experts. On the other hand, the experts understood "natural period," but they could not fully outline the concept to the volunteer supporters. The volunteer supporters learned something they had not previously known through mutual communication with the experts. Through learning about the existence of the concept of a natural period, the volunteer supporters knew what they did not know before and represented their understanding of the concept by making a miniature pendulum.

An example of tacit knowledge is the concept of a "natural period" that appeared in the mutual communication between the experts and volunteer supporters mentioned above. The volunteer supporters found the concept of a natural period to be something they did not know, so they asked the experts.

By explicitly indicating tacit knowledge, the volunteer supporters got a chance to learn something new. At the same time, the volunteer supporters gained an opportunity to follow or copy remarks or behaviors from the experts' tacit knowledge. On the other hand, the experts had a chance to rethink a remark or behavior that they regarded as too common or natural to require any additional explanation. At the same time, they had the opportunity to reconsider and adjust remarks or behaviors underlying their own framework of knowledge as an expert in accordance with the needs of the volunteer supporters in leading a guided tour. In other words, they renewed the concept of a natural period through copying and arranging remarks or behaviors according to the needs of the visitors. To put it plainly, they learned with each other.

The situation above indicates that the volunteer supporters performed active learning through their role and identity by participating in the community of practice at Abuyama Open Laboratory.

This leads to a further consideration of mutual negotiation between experts and volunteer supporters. In the following section, we consider this relationship from an expert's point of view.

5.2 *Change in Identity of Experts*

In this section, we consider the change in identity of an expert (Iio). The first author interviewed Iio on January 24, 2013. The interview provided good evidence to show the change in identity.

According to the previously mentioned theoretical assumptions, we consider identity to be a result of change in which a full participant (Iio) comes to accept "constructively naïve perspectives or questions" that new participants (volunteer supporters) bring in from outside the community of practice of Abuyama Open Laboratory in the following form: "Inexperience is an asset to be exploited."

In the interview, the author asked Iio the following question:

> Through conducting Abuyama Open Laboratory, has your view on your research changed?

Iio said to the author in response:

> Even when explaining old seismometers that I no longer use for seismic observation, volunteer supporters often ask me questions about how the item relates to useful outputs. Such interactions make me feel that I have to output good results through my seismology research activities.
>
> When I explain my research to people who are unfamiliar with it, I interpret it for myself beforehand because I have to confirm or understand basic knowledge again before explaining it to others. It is necessary for experts to repeatedly check basic knowledge that they usually regard as obvious, and to understand it again. Therefore, the Abuyama Open Laboratory gives me valuable and useful chances.

The author went on to ask to Iio:

> Through conducting Abuyama Open Laboratory, have there been any changes in your self-perception?

Iio answered as follows:

> My seismology work covers just a small area because seismology covers various fields and has been built on the culmination of predecessors' persistent efforts. Consequently, a lot of knowledge needs to be accumulated before understanding and proceeding with my research.
>
> It is difficult to explain my research to the general public in cases such as Abuyama Open Laboratory or similar public lectures. Metaphorically speaking, let us suppose that my research is a stairway that has 10 steps. I assume that the general public understands only the first step because they don't have basic knowledge of seismology, so I start my explanation from the beginning level. On the other hand, I have to give my lecture within a specified time. By contrast, I can make experts understand all 10 steps because I assume that they not only have basic knowledge but also know up-to-date technical terms through continuous seismology research activities. However, I'm afraid that the stairway is cut off; that is to say, I don't think experts always have a clear understanding of basic seismology knowledge. Basic seismology knowledge and the latest technical terms are not always shared. I therefore think it is necessary for seismology experts, including myself, to review basic seismology knowledge usually regarded as obvious and to understand it again from time to time.

Considering the interview above, Iio's change in identity can be explained by LPP theory as follows: First, Iio, who is theoretically a full participant of the community of practice at Abuyama Open Laboratory, recognized new participants and gradually came to acknowledge them as volunteer supporters who were theoretically full participants. Second, Iio interacted with volunteer supporters concerning guided tours, which theoretically is a central practice of Abuyama Open Laboratory. Third, through this interaction, Iio confirmed his framework of seismological knowledge and rebalanced his behavior accordingly. He also rebalanced his behavior with the intentions of the volunteer supporters; namely, he refined his role in the community of practice at Abuyama Open Laboratory, which can be explained as a change in identity.

5.3 Change in Identity of Pupils

In this section, we consider the result of learning through the participation of pupils in the latest seismic observation project. It must first be acknowledged that changes in the identity of pupils emerged less clearly compared with the volunteer supporters in the Abuyama Open Laboratory.

First, as of July 2014, more than 100 pupils of the two previously mentioned elementary schools participated in the study program to practice seismometer maintenance, which lasted 4.5 years. While the facts above show that the study program allowed pupils to become involved in seismometer maintenance as an elementary part of seismic observation, it is hard to say whether the pupils refined their role in maintenance of the seismometer voluntarily or independently to improve seismic observation activities or improve their skills.

On the other hand, it is likely not far from the truth that the pupils were conscious of their role in seismometer maintenance because seismic observation was considered important. For example, such an attitude appears in the following remark: "We learned not to play around near the Manten seismometer in order to avoid having extra noises recorded by the machine." Their behavior in making a box covering and protecting the Manten seismometer is another example of the same point.

The most likely explanation is as follows. According to LPP theory, participating in a community of learning where knowledge transfer is not central to learning, an "ideal" central participant (Fig. 5) does not respond to an expert, as argued in the deficiency model of risk communication theory. In other words, the central participant is

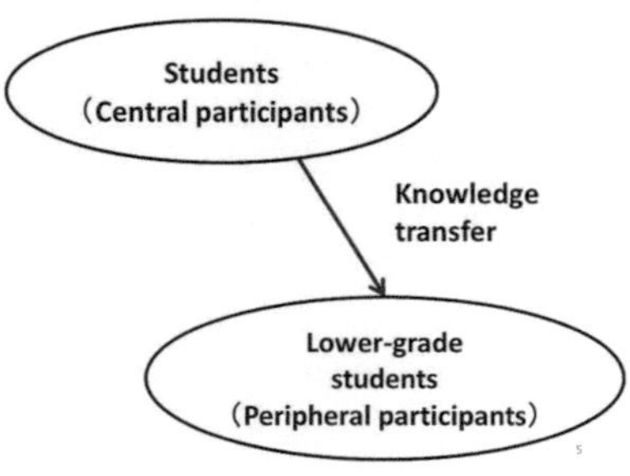

Fig. 5 Knowledge transfer from students to lower grade students

not opposed to learning. Participation in the "community of practice," where knowledge is shared rather than transmitted in one direction from experts to non-experts, becomes a catalyst for shared learning.

This theoretical application could be significant for the following two reasons. First, it is possible to assume that ideal and abstract "central participants," based on the "deficiency model," are set against real and concrete "full participants," based on LPP theory, and this provides a continuous opportunity to induce dialogue or communication concerning what knowledge or techniques should be handed down because it is easier for new participants, who are new to LPP theory, to share the concept of "central participant" than that of "full participant." Another possibility is that, from a theoretical perspective, the "deficiency model," which has a negative meaning, or is at least criticized in the context of science communication theory, took on a positive and constructive meaning in the context of LPP theory. Corresponding to the general concept of "expert," the concept of "central participant," which does not exist but can be thought of as an object, can be all the more important.

Regarding the practical importance of this theoretical application, even if certain knowledge is found to be valid and effective for disaster prevention activities, it is costly for the receiver to learn and master it. Therefore, it is reasonable to carefully consider whether to hand down the knowledge or technique before transmitting and diffusing it. From this, the "deficiency model" can be thought of as important for this discussion. Since what knowledge or technique to hand down could be a common precondition of discussion, any miscommunication could be potentially resolved by transferring knowledge and techniques.

In addition to the cost of transferring knowledge and techniques, this theoretical application could be effective in solving the problem of pessimism and optimism, as stated previously.

LPP theory should therefore be concluded to be a good platform for bridging the communication gap between experts and non-experts.

5.4 Implication for Open Science Movement

Finally, I would like to highlight the aspects of the research discussed in this chapter that merit attention from the perspective of open science, a concept that is drawing global interest and is being promoted by the Japanese government as a pillar of its comprehensive science and technology policy. The open science movement (Nielsen 2013) aims to enable a wide range of people—scientists as well as members of the public—to participate in scientific research in a variety of ways, including by accessing the results of scientific research and related information. In other words, the aim is to reform science so that is open to society.

Narrowly defined, the term open science sometimes refers to making the data and findings that form the foundation of scientific research accessible to more people through open access to academic databases. The term can also include conventional science communication, that is, communication and outreach by scientists intended to make scientific topics more easily comprehensible to the general public.

However, the main task of the open science movement is to reconsider the roles of scientists and non-scientists, moving away from a system in which the former produces and presents scientific data and findings to the latter and instead opening the process of science itself to the public. In other words, it is essentially an effort to reorganize science itself so that it becomes a collaborative enterprise involving a wide range of people, including members of the general public who are not scientists.

Initiatives toward this end have already gained significant public support in some fields. For example, amateurs have long made contributions to astronomy, and these non-scientists have in fact discovered a number of new stars. In the field of biological observation, data from citizen volunteers have pinpointed the behavior patterns and habitat range of certain plants and animals, helping to support scientific research on those species. In the humanities, non-experts have helped to collect, organize, and interpret old documents for historical research. Unfortunately, this type of collaboration has thus far played a minimal role in the field of disaster prevention and mitigation discussed in this chapter.

Against this context, the examples highlighted in these pages can be viewed as attempts to incorporate the philosophy of open science into disaster prevention and mitigation studies and use it to trigger fundamental changes in education and outreach on that topic. For example, the volunteer staff at the science museum project at Abuyama Observatory were at first engaged primarily in interacting with visitors to the museum (for instance, explaining the seismographs), and in that sense were limited to conducting outreach in the narrow sense of the term. However, they later began working under the direction of experts to observe aftershocks of local earthquakes and organize data, thereby participating deeply in the advancement of the core scientific undertakings of data collection and interpretation. In addition, although the Manten Seismic Observation Project is still a work in progress, elementary students and teachers are contributing indirectly and partially to the advancement of science by helping maintain seismographs installed at elementary schools.

By using the language introduced in this chapter, the essence of open science can be expressed as follows. Open science is an effort to reorganize the community of practice that supports the societal activity of science; in particular, it is an effort to free scientists (experts) and citizens (non-experts) from their conventional identities so that they may gain new identities. The initiatives described in this chapter represent attempts to bring open science to the hitherto unreformed field of disaster prevention and mitigation, where tall, thick walls have long separated experts and non-experts.

These initiatives of course represent only the first steps toward this goal. The effectiveness of such efforts must be proven through many more similar projects. There is ample reason to believe that will happen. In recent years, initiatives that share the same fundamental vision as open science have increased in the field of disaster prevention and reduction. One such example is the attempt to use information collected by citizens (non-experts) to improve the resolution of data on localized, fast-changing weather disasters (such as localized heavy rains and winds or sudden surges in river levels) that cannot be adequately captured by observation networks established by experts (Takenouchi et al. 2014). Citizens are also collaborating in efforts to study the history of recurring disasters such as earthquakes and tsunamis, particularly those for which modern observational data do not exist, by interpreting old documents (Kano 2017). These initiatives suggest that communities of practice based on dichotomous relationships between experts and non-experts are being reorganized within the framework of open science, with both experts and non-experts beginning to take on new identities, and that there is an excellent outlook for the future of such efforts.

The concept of reorganizing the communities of practice that form the basis for science—in other words, opening up science to the public—goes a step beyond the frameworks that have been proposed again and again, such as fusing the humanities and sciences, promoting interdisciplinary research, and emphasizing the importance of both natural and human sciences when conducting disaster prevention education. This is not merely an effort to integrate or reorganize various disciplines within academia but rather to change the position of academic endeavors within society as a whole. To date, research conducted by scientists has existed as a separate strand within society from grassroots or community disaster prevention and mitigation activities conducted by ordinary citizens. By using open science as a means to unify these strands, they are being repositioned as a single movement within society.

6 Conclusion

In the community of practice of Abuyama Open Laboratory, experts (full participants) and volunteer supporters (new participants) maintained ongoing mutual negotiations concerning guided tours of the observatory. After this initial stage, the cross-boundary collaboration has developed into the secondary stage in which volunteer supporters are deeply involved in a seismic observation activity itself in the recent local earthquake. This resulted in volunteer supporters taking on a new identity and experts rediscovering the fundamentals of their role.

In the community of practice of the collaborative study program between the elementary school and the Manten Seismic Observation Project, a constructive meaning was identified through an LPP analysis of the apparently passive attitude of the students. Rather than finding the deficiency model flawed as a cause of the communication gap, we found that it served as a communication bridge between experts and non-experts.

These findings suggest that LPP theory allows disaster education to be recognized and discussed in the context of a specific and substantive "community of practice." We also pointed out that our initiatives are closely related with open science, because open science is an effort to try to reorganize the community of practice, in other words, an effort to free scientists (experts) and citizens (non-experts) from their conventional identities so that they may gain new identities.

References

Abuyama Open Museum Project (2011) Leaflet of first special open house for public information (not published)

Fujigaki Y, Hirono Y (2007) Science communication theory. University of Tokyo Press, Tokyo (in Japanese)

http://www.tsunamachimitakai.com/pen/04_003.html. Accessed 30 July 2016

Iwahori T, Yamori K, Miyamoto Y, Shiroshita H, Iio H (2017) Disaster education based on legitimate peripheral participation theory: a new model of disaster science communication. J Nat Disaster Sci 38(1):1–15

Kano M (2017) Outline of Citizen's challenge for decrypting old documents about historical earthquakes. Rep Kasama 63:53–56 (in Japansese)

Kumamoto K (2011) The great East Japan Earthquake in reality is not "unexpected." Tsunamachi Mitakai Club (in Japanese)

Lave J, Wenger E (1991) Situated learning: legitimate peripheral participation (Japanese version: translated by Saeki, Y. Sangyo Tosyo 1993). Cambridge University Press, Cambridge

Nielsen MA (2013) Reinventing discovery: the new era of networked science. Princeton Univ Press

Nogami T, Yoshida F (2014) Disaster myths after the great East Japan disaster and the effects of information sources on belief in such myths. Disasters 38(2):190–205

Okada N, Tao Y, Kajitani Y, Shi P, Tatano J (2011) The 2011 Eastern Japan great earthquake disaster: overview and comments. Int J Disaster Risk Sci 2(1):34–42

Shiroshita H, Yamori K (2011) Participatory disaster management learning built on the theory of legitimate peripheral participation. J Disaster Res 6(2):258–270

Takagi K (1999) Identity construction in legitimate peripheral participation theory. Bulletin of the Center for Educations of Children Overseas Tokyo Gakugei University 10:1–14 (in Japanese)

Takenouchi K, Kawata Y, Nakanishi C, Yamori K (2014) Collaboration on local weather information between weather forecasters and weather information users. J Nat Disaster Sci 35:67–80

The Japan Times (2004) Derailment mars shinkansen safety myth (Nov. 19, 2004) http://www.japantimes.co.jp/news/2004/11/19/national/derailment-mars-shinkansen-safety-myth/#.WEZjhtH_q74. Accessed 6 Dec 2016

Yamori K (2012) Improving disaster risk communication: a paradigm shift in disaster information research. Minerva Shobo, Kyoto (in Japanese)

Yamori K, Suwa S, Funaki N (2007) Dreaming disaster education. Koyo Shobo, Kyoto (in Japanese)

Do Developed Countries Learn DRR from Developing Countries?

Hideyuki Shiroshita

Abstract The international community's strategy to assist Nepal in recovering from the 2015 Nepal Gorkha earthquake will be considered from a disaster educational perspective. In the first half of the chapter, the importance of 'unfixing' the relationship between experts on DRR and lay people will first be discussed since this relationship provides a platform for the conventional idea of disaster education, that is, knowledge transmission from experts to lay people. In order to unfix the relationship regarding disaster education, the possibility of mutual communication in the recovery process of the 2015 Nepal Gorkha Earthquake will be discussed. In particular, we attempted to discover the perceptions and indigenous knowledge of those affected local people in Nepal as transferring knowledge from those who are not supposed to have scientific knowledge is an important part of mutual communication. Based on a series of interviews, as of 2019, it is difficult to clearly point out what are the local knowledge or skills. However, it is possible to propose a hypothesis that there have been rules of thumb that have emerged in the recovery process that have guided the survivors in their efforts to recover from the disaster.

Keywords 2015 Nepal earthquake · Education · Learning · Rules of thumb · Local knowledge

1 2015 Nepal Gorkha Earthquake

It has been said that every 70–80 years, in Nepal, in particular in the central part of the country that includes its capital city Kathmandu, a huge earthquake occurs. This is referred to as '70-year rule' in Nepal (Oven et al. 2016). The last huge earthquake occurred in 2015, namely, the 2015 Nepal Gorkha earthquake, and it killed nearly 9,000 people in total. Some natural disaster experts are aware of the risk of earthquakes in the country and are making an effort to get people prepared for future earthquakes.

H. Shiroshita (✉)
Kansai University, Takatsuki, Japan
e-mail: hideyuki@kansai-u.ac.jp

K. Yamori (ed.), *Disaster Risk Communication*, Integrated Disaster Risk Management,
https://doi.org/10.1007/978-981-13-2318-8_7

1.1 An Overview of Nepal

The Federal Democratic Republic of Nepal (hereafter Nepal) is a landlocked country in South Asia. Its territory covers an area of 147,181 km^2 with a population of 26,494,504 as of 2011 (Intensive Study and Research Center 2013). The shape of the country stretches from east to west, and its length is approximately 800 km. On the other hand, the length from north to south is around 200 km. The lowest altitude is 90 m, and the highest altitude is 8848 m, which is the peak of Mount Everest (Intensive Study and Research Center 2013). The country is famous for the Himalayan mountains and several historical heritage sites including the Kathmandu Valley, a UNESCO World Heritage site. The country attracts many tourists from all over the world, and tourism is one of its main industries.

However, if we evaluate the country using several commonly used indexes, such as GDP, life expectancy and university enrolment ratio, Nepal is ranked low in all these indexes. For example, its per capita GDP is 729.1 USD as of 2016, which is approximately 1/80 that of the USA's GDP of 57,638.2 USD in the same year (The World Bank Data 2016). In addition, the country suffered a civil war from 1996 to 2006, which also adversely affected the country's development. This is why the country is described as 'one of the world's poorest countries' (BBC News 2018).

In addition, the country is located on the Alpine–Himalayan belt, where many earthquakes, including huge ones, often occur. While there are no typhoons or hurricanes in Nepal, they often have heavy rains in the summer. Thus, floods and landslides frequently occur. Moreover, snow-capped mountains bring avalanches. These hazards have brought severe damage to the country. Table 1 shows the number of people affected by the hazards in the last 5 years.

Nepal is affected by many natural hazards, and the interaction between these hazards and the vulnerable society bring huge losses to the country.

Table 1 Number of dead and missing persons on account of major hazards in Nepal from 2013 to 2017

	2013	2014	2015	2016	2017
Landslides	109	242	137	155	85
Floods	264	262	3	137	208
Earthquakes	0	0	8962	0	0
Thunderbolts	147	97	81	106	85
Avalanches	16	19	2	0	2

Source Nepal disaster risk reduction portal (Government of Nepal 2018)

1.2 2015 Gurkha Earthquake

At 11:56 am on 25th April 2015, an earthquake of magnitude 7.8 occurred approximately 80 km west of Kathmandu. The biggest aftershock of magnitude 7.3 occurred on 12th May. These two earthquakes caused severe damage to the country and even to surrounding countries. Only focusing on Nepal, the earthquake killed 8,962 people (Government of Nepal 2018). In addition, 773,095 buildings were completely damaged, while 298,998 buildings were partially damaged (Government of Nepal 2018). Many heritage sites including Bhaktapur, Kathmandu, and Patan Durbar Squares were also damaged. The Dharahara aka Bhimsen Tower completely collapsed. The tower had been earlier broken in the 1934 earthquake and rebuilt thereafter. It was destroyed again in 2015. However, reconstruction of the tower started in 2017. Moreover, many temples including Swayambhunath and Boudhanath were also severely damaged.

This damage to heritage sites resulted in indirect damage to the society since tourism is one of the biggest industries of the countries as stated above. Regarding the number of tourists, 790,118 people visited Nepal in 2014; however, the number decreased to 538,970 in 2015 (Ministry of Culture, Tourism & Civil Aviation 2016).

It was anticipated that the central part of the country would experience an earthquake since there have been periodical earthquakes every 70–80 years (Oven et al. 2016). The last major earthquake before the 2015 one occurred in 1934. However, although anticipated, the country's preparedness was insufficient for the earthquake.

1.3 Response and Recovery

Right after the occurrence of the earthquake, a meeting of the Central Natural Disaster Relief Committee (CNDRC) was held in accordance with the Disaster Relief Act 1982. CNDRC recommended to the government that the most severely affected 14 districts (Fig. 1) be declared an emergency area.

The meeting also decided to establish a Response Coordination Centre to accelerate the search and rescue operations. As of the 5th of May, 4,050 people from 76 search and rescue teams and 135 canines had been mobilised from 34 different countries (Ministry of Home Affairs 2015). As stated above, since a large number of the buildings had been destroyed, CNDRC decided to give priority to providing shelter, food, and medicine on the 3rd of May.

Around one month later, the government started to distribute 15,000 rupees to each family who had lost their house for building a temporal shelter. A group proposed a method of building a shelter within 15,000 rupees. The proposed shelter, called a tunnel shelter, consists of three or four arch-shaped rebars and a corrugated tin metal plate roof as shown Fig. 2. However, both sides should be constructed by the owner.

After the earthquake, many forms of assistance by foreign governments as well as INGOs were declared and implemented. For example, the International Federation

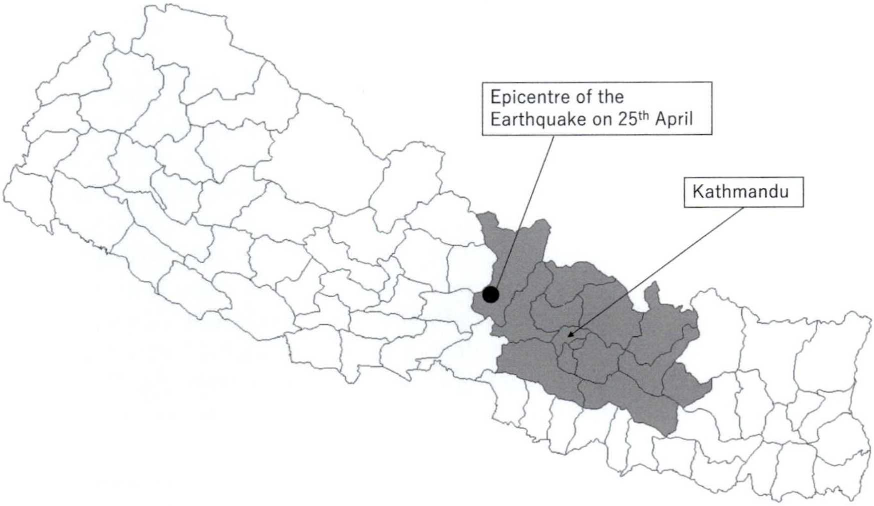

Fig. 1 Map of the epicentre and the most severely affected districts declared by the central government. *Note* The most severely affected districts (shown in grey) include Gorkha, Sindhupalchowk, Dhading, Kavre, Dolakha, Nuwakot, Ramechhap, Sindhuli, Rasuwa, Kathmandu, Lalitpur, Bhaktapur, Makwanpur and Okhaldhung. *Source* of the original political map: Obtained from Wikimedia Commons as a public domain map

Fig. 2 Photo of the tunnel shelter

of Red Cross and Crescent Societies provided immediate medical assistance. They provided temporal hospitals and clinics in the affected areas. As of 2018, the total amount of external support including pledges was 4,112 million USD. The number of projects including pledges exceeded 300 (Ministry of Finance 2018).

However, it is necessary to revisit the question as to what 'build back better' is. As the World Disasters Report 2014 points out, 'Those organizations (note: organizations that are involved in disaster risk reduction) often assume that people want to take action about serious hazards, using the same logic that they use (International Federation of Red Cross and Red Crescent Societies 2014)', outsiders consciously or unconsciously use their logic, and they apply their sense of values for DRR when they assist the people. However, in Nepal, in particular the rural areas of the country, it is difficult to implement the same way of recovering as the one, for example, adopted by Japan after the 2011 Tohoku earthquake. In other words, the resources for recovery are very limited in the rural areas. This fact is appreciated by the government. For instance, at the reconstruction conference, one suggested utilising 'rules of thumb' in the rural areas while applying 'science' in the urban areas (Thiruppugazh 2015). It should be carefully considered what 'build back better' for the country is.

Another example of the importance of thinking the country's needs can be argued from cases of failure of assistance. As Sato (2005) argues, lack of careful consideration about development aid could bring negative effects to the recipients. The author also saw several cases of failure of assistance in Nepal. Most of the causes of the failure arose from not thinking about what happens after the termination of the assistance. For example, a set of PA systems for evacuation had been installed in a village and had broken down after a certain period of time. However, since no one could repair it owing to the lack of resources, the system was left as it is. Another example is that a rain gauge was installed in a local school for precipitation data collection; however, after running out of the provided paper for recording, they could not get additional supplied of the special paper for the recorder.

In this light, this chapter explores how the international society's strategy to assist Nepal in recovering from the earthquake is considered and proposed. Although it can be discussed from several aspects of the assistance, the approach from learning is taken. This is because learning can be defined as a whole process of daily life.

2 What Is DRR?: In Developed Countries and Developing Countries

In UNISDR's definition of DRR, 'The concept and practice of reducing disaster risks through systematic efforts to analyse and manage the causal factors of disasters' (UNISDR 2012), it does not appear that DRR includes the practices adopted after disasters. In other words, response, recovery and reconstruction after disasters are not included in the definition of DRR. However, as a disaster management cycle targets not only actions before a disaster but also actions after the disaster, DRR should be understood as a permanent practice for disasters. The Japanese-language term for DRR is 'bousai', and the meaning of the word includes response, recovery, and reconstruction. It appears that a disaster breaks time into two phases, before and after; however, real time has not been broken. It can be said that it is impossible and meaningless to discuss which part of the practices is DRR and which part is not. In this light, DRR is defined as a whole practice against disasters in this chapter.

2.1 Why Has DRR Been Insufficient?

It goes without saying that DRR is important for the international society. In addition, as the Sendai Framework states, DRR requires an all-of-society engagement and partnership (United Nations 2015), and all people are requested to participate in its practices to realise effective DRR. It is, however, difficult to achieve this goal.

Japan is one of the most disaster-prone countries and has recently experienced several catastrophes such as the 2011 Tohoku earthquake and tsunami and the 2015 Kumamoto earthquake. However, as several preceding research studies have shown, the Japanese recognise the risk and have knowledge and skill in dealing with it, but they do not prepare and behave properly right after a disaster (Shiroshita 2018). For instance, it is strongly recommended that furniture be secured to a wall or pillar in order to prevent it from falling down when an earthquake occurs. However, the latest national poll, which was taken after the 2011 Tohoku earthquake showed that around 60% of responders had not secured the furniture. In addition, the top response out of 13 reasons why they do not secure the furniture is, 'I want to do it, but it has been postponed'. The reason related to lack of knowledge or skills is ranked ninth. Thus, it is not the main reason for not securing the furniture. Another example is from the 2011 Tohoku earthquake. Since the 1970s, the western part of Japan has prepared for an earthquake and tsunami as the prediction that a big earthquake and tsunami would occur had come to light in 1976. However, when the Tohoku earthquake occurred, only 2.5% of the residents were ordered to be evacuated to the designated evacuation place (Shiroshita 2014). In short, people understand the risk and know how to prepare for and deal with it, but they do not.

However, it is not appropriate to apply this reason in the case of developing countries. The lack of opportunities to learn about disasters is the main reason. For example, research showed the deficiencies and limitations in the process of DRR knowledge dissemination among the local communities of Nepal (Tuladhar et al. 2015). However, the government of Nepal has already incorporated disaster management subjects in school curricula (Ministry of Home Affairs 2016). This contradictory situation showed that even when the government establishes a rule or system, it takes time to spread over the country. In addition, as mentioned above, the situation is worse in Nepal as it is interconnected to poverty. Since poverty is one of the obstacles to the spread of education, international cooperation would be a great help in spreading education. This could be applied to a certain extent. For instance, when the Indian Ocean Tsunami occurred in 2004, several types of research showed that the lack of knowledge about tsunami was one of the causes of the expansion of damage, in particular, casualties. However, there is no guarantee that the more knowledge people have, the more prepared they will be. This is because the reason that people in Japan do not prepare for disasters is not lack of knowledge about DRR.

Both in Japan and Nepal, people do not prepare and behave properly immediately after a disaster. However, people in Nepal faced a comparatively more severe situation than Japanese people. For instance, the annual death toll of natural disasters in Nepal is usually more than that of Japan although the population of Japan is four times more than that of Nepal. The difference between these countries is based on the level of scientific countermeasures installed mainly by experts on DRR. In Japan, there are countermeasures such as a shake-proof building code, earthquake early warning (EEW) system, and gas meters that automatically cut off when an earthquake occurs. In other words, even among people who do not prepare for disasters, these countermeasures installed by the experts prevent or reduce the damage from disasters up to a certain degree. On the other hand, people in Nepal must prepare and deal with disasters by themselves as there are very limited scientific countermeasures available in the country.

As Table 2 shows, once scientific countermeasures have been installed, people neglect to prepare for disasters by themselves. The relation between the installation of scientific countermeasures and people's preparation is a trade-off. It is nonsense that doing away with scientific countermeasures could make people more involved in DRR. Instead, we should explore ways of including both scientific countermeasures and people's involvement.

There are several preceding types of research on how to assist in the development of DRR in Nepal. Of course, these types of research show an understanding of the socio-economic differences between Nepal and the country of the authors. However, most of these try to implement similar methods of DRR that have worked in the country of origin of the authors; only a few appreciate the implementation in Nepal. In other words, many forms of assistance in the development of DRR in Nepal consist of transferring scientific countermeasures from donor countries faced the problem that people do not prepare for disasters by themselves. It is necessary to recognise that the current situation regarding DRR in Nepal is not tragic. Discussing how to maintain people's attitudes in Nepal is important.

Table 2 Comparison of the status of DRR between Nepal and Japan

	Scientific countermeasures	Knowledge level of DRR	Who mainly prepare for disasters?	Are people prepared?
Nepal	No (limited)	Low	Lay people	Yes (limited)
Japan	Yes	High	Experts	No

2.2 Are People in Nepal not Prepared for Disasters?

Nepal is usually understood as a country that is not prepared for disasters because of poverty. As Laursen stated, very few have begun preparations in Nepal (Laursen 2015). However, looking at the society carefully, several examples of preparedness against disasters can be found. For instance, most of the houses have a huge water tank on the rooftop because the water supply is suddenly cut off without any notice. If they reserve water instead of using the water supply directly, then even if the water supply is cut, they can still have water for a few days. In addition, when we visit the houses in Nepal, we can find either a big battery, usually that of a vehicle, or a power generator. This is for load shedding that is sometimes unplanned.

In Japan, the government provides the residents with suggestions on how to prepare for disasters. These suggestions include packing an emergency 'go-bag', taking preventive measures to avoid falling furniture, building retrofitting, evacuation drills, etc. However, most Japanese people are not well prepared. It can be said that more Nepali people are prepared than Japanese people if we evaluate the preparedness using Japanese standards, in particular, the emergency go-bag. This is because, the government suggested putting a few bottles of water, flashlight and a first aid kit in the bag. In this light, people in Nepal have already made preparations for water and light in their daily life. Interestingly, people in Nepal do not consider that they are already prepared for disasters, and they think they cannot prepare for it because of their poverty. If they do not recognise this fact, they will lose this good DRR practice, that is, preparing for water cuts and power cuts by themselves in the near future, since developing the country with science-based technological measures will make it disappear. In other words, the preparedness will be a special activity only for disasters, and it will be separated from their daily life. Not only implementing scientific countermeasures of DRR but also keeping and utilising the wisdom of living is also important for realising viable DRR in Nepal.

3 Transformation of DRR and Disaster Education

3.1 Transformation of DRR

The definition of contemporary DRR normally relies on the assumption that there is a special relationship between natural phenomena and people. This relationship defines DRR as the use by people of sciences such as seismology, meteorology and volcanology to observe and understand the natural phenomena and use of fields such as architectonics, civil engineering and urban planning to protect the people from extreme natural events (Fig. 3).

This contemporary definition is not a universal one. However, most people, mainly those in developed countries, accept this definition. For these people, disasters appear to be failures of science or science-based technology. For instance, if people cannot

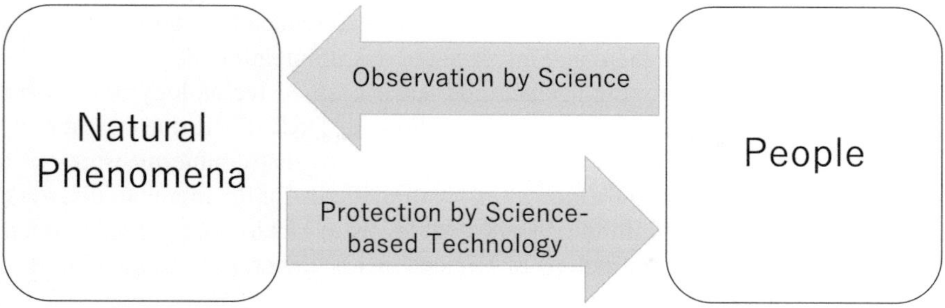

Fig. 3 Public image of the relationship between natural phenomena and people

evacuate in time because they do not get an early warning, this situation is understood as a failure of the information system. If a building collapses because of an earthquake, this is understood as a failure of the construction or building code. In short, earthquakes do not kill people; rather, it is the buildings that kill people.

However, as stated, this is based on one of the special relationships between natural phenomena and people. If people do not put science and science-based technology between natural phenomena and people, the understanding of disasters is not the same. World Disasters Report 2014 shows several interesting cases (International Federation of Red Cross and Red Crescent Societies 2014). In West Africa, a volcanic eruption is understood as God's anger; thus, local people believe that they can calm the God's anger by making a sacrifice. Even in the US, several politicians and even people that were affected by Hurricane Katrina believed it was God's punishment for sinful behaviour in New Orleans. It is possible to say that their understanding is not scientific; however, it is impossible to say that these are wrong as these are people's perceptions.

The purpose of this chapter is not changing the perception of those who are not scientific but changing ideas of those who are scientific. Of course, this does not mean avoiding science and science-based technology. Instead, it means changing the perception of science and science-based technology from being the only option for DRR to being one of the options for DRR. In other words, options other than science and science-based technology for DRR will be visualised. Moreover, DRR based on science and science-based technology should be understood as a result of choice.

Science and science-based technology for DRR have a significant impact on DRR. Several data show that the implementation of scientific and technological countermeasures has reduced human loss by natural hazards (Shiroshita 2012). However, as Yamori and Onishi stated, if people rely too much on science and science-based technology, then the side effects of those become major (Yamori and Onishi 2016). Strictly speaking, the relationship between science and science-based technology and people is pointed out as the cause of this side effect. For example, if a huge levee is constructed, the residents of that area think evacuation is not needed. In addition, if they believe they can get proper evacuation information in a timely manner, they will not evacuate from a risky location to a safer place enough in advance. Although both the levee and the evacuation information are meant to reduce disaster risks, people begin to rely too much on them and increase the risk unintentionally.

As stated above, improving science and science-based technology is essential for DRR. However, these are not the only options of DRR. DRR does not merely entail setting up scientific and science-based technological countermeasures in a society. Rather, DRR should include a variety of methods for dealing with disasters. This expansion of the definition of DRR enables us to tackle both the side effects mentioned above and the limitations of science, that is, uncertainty and ambiguity.

3.2 Transformation of DRR

Disaster education is usually understood as a transmission of knowledge from experts on DRR to non-experts. For example, UNESCO defines disaster education as 'education for disaster preparedness can provide life-saving and life-sustaining information and skills that protect in particular children and young people during and after emergencies. (*snip*) UNESCO provides tools for integrating Disaster Risk Reduction into schools and provides training for teachers and educators on how to educate children and youth to prepare for disasters' (UNESCO 2012). The point of this conventional definition of education is that the transmission is one-way, namely, those who have it to those who do not have it. Thus, in the above-mentioned definition, the experts can be replaced by teachers or adults and the non-experts can be replaced by students or children. Disaster education is usually based on the fixed relationship between those who teach and those who are taught. This is because as Preston (2012) points out, there are a very limited number of preceding studies on disaster education. Preston also argues the transfer of knowledge is a part of disaster education. However, this limited definition and the aforementioned fixed relationship are widely accepted. This fact regarding education makes it difficult to change the idea of DRR since most scientific and science-based technological countermeasures are implemented by the experts. Keeping this fixed relationship makes people believe that DRR is implemented by the experts as they are the only people who understand and utilise science and technology. In order to change the fixed idea of DRR, disaster education also needs to be expanded. Since DRR should include a variety of methods for dealing with disasters, disaster education also should be a mutual method of communication in order not only to transfer scientific ideas on DRR but also to transfer rules of thumb. In other words, disaster education should not be defined as giving a meaning of DRR from those who have to those who have not. Instead, it should be defined as a co-creation of the meaning of DRR through a mutual form of communication, namely, co-learning.

4 What We Can Learn from the Affected Area of the 2015 Earthquake

As mentioned earlier, since earthquakes in Nepal occur periodically, disaster education is necessary for not only recovery and reconstruction but also mitigation and preparedness for a future earthquake. Delivering a lecture or conducting training for those who are affected is one of the options of disaster education. It is easy to find this type of disaster education practice in Nepal. However, this type of narrowly defined disaster education, namely, the transferring of scientific knowledge and skill, has several problems. The first problem is that even the cutting edge of scientific knowledge is a transferred implementation that would be difficult because of poverty. The second problem is that even if scientific knowledge is utilised, the scientific knowledge includes uncertainty and ambiguity. The third problem is the fixed relationship between those who have knowledge and skill and those who do not. This fixed relationship lets non-experts rely on the experts too much. In order to avoid these problems, disaster education should be understood broadly. In particular, the fixed relationship should be unfixed. As mentioned above, to realise mutual communication it is necessary that those who are considered 'have-nots' should transfer knowledge or skills to those who have.

A series of interview surveys has been undertaken in Sindhupalchok District in 2016, 2017, and 2018. Sindhupalchok District is one of the most affected districts in the country. It had a population of 287,798 in 2011, but 3,570 people were killed by the earthquake (Government of Nepal 2018). If we compare this rate to the 1995 Kobe earthquake, Kobe had a population of 1,518,982 in 1994, and the earthquake killed 4,564 people (Hyogo Prefectural Government 2005). The death rate (casualties/population) of the former earthquake is approximately four times higher than the latter. Sindhupalchok District has 78 towns and villages.[1] The interview surveys took place in Chautara Municipality (in 2016, 2017. and 2018) and Melamchi town (in 2016 and 2018).

4.1 *Interview Survey in 2016*

The interview took place in the middle of March. Interviewees were randomly (strictly, it was snowball sampling) selected as it was difficult to conduct a well-organised interview survey in the affected area. We visited the site and found people there and asked for their cooperation in the survey. As the author does not speak Nepali, a Nepali-English speaker was hired as an interpreter. In Chautara, eight persons participated in the survey and ten persons participated in Melamchi.

The interview was semi-structured and mainly asked about how to deal with the earthquake and the preparedness before and after the earthquake. As for the

[1] After the earthquake, although not because of it, the governance system in Nepal has drastically changed. These were not towns and villages but Village Development Committees (VDCs).

preparedness before the earthquake, the question was 'Had you prepared for earthquakes before the earthquakes?' All 18 interviewees answered no. The reasons for unpreparedness were roughly classified into two patterns. The first was they had not received any information regarding earthquakes in advance. Thus, they did not prepare for earthquakes. The other reason was that they had received information about earthquakes and how to deal with them; however, they did not know when an earthquake would occur. Thus, they did not prepare for it. The priority of preparedness for earthquakes was low.

Regarding the preparedness after the earthquake, the question was 'Have you prepared for future earthquakes after the earthquakes?' More than half of the interviewees (13 out of 18) answered yes. Most of the interviewees mentioned how to reconstruct their house. This included the information that they should build a single-story house, avoid using stones for the wall, and use a light roof such as a tin plate. They have learned and found that the buildings that collapsed used heavy materials such as stones and vulnerable structures are those that use mud instead of cement.

Although they had neither disaster education nor training, they have learned from their experience and tried to implement these learnings.

4.2 Interview Survey in 2017

In the middle of March, the second interview survey was conducted only in Chautara. We tried to conduct interviews with the same interviewees as in 2016. However, four out of eight interviewees were not available at that time. Two interviewees were absent; as for the other two interviewees, we spoke to their family members. In total, we received six responses this time. The style of the interview was the same as in 2016, that is, semi-structured.

It has been nearly 2 years since the earthquake. However, it seemed the speed of reconstruction was slow. Only one interviewee had built a temporal house; others who lost their houses were still in the temporal shelter. Most of the interviewees complained about the government's method of distributing public donations. Some interviewees had a problem with receiving the donation, and they were trying to solve the problem. It seemed that all the people were frustrated because of the slow speed of the reconstruction.

We asked about preparedness after the earthquake again; the question was 'Have you prepared for future earthquakes since the earthquake?' One interviewee answered that he started to teach children about DRR. The other said he decided to build a safe house (single-story and tin roof) and started to reserve some food in case of a disaster. In addition, a lady answered that she had found that even among similar RCC buildings, some had collapsed but others had not. She had surveyed on her own and found the difference between these buildings was the foundation of the buildings. She learned that it is important to have a proper foundation to construct a building strong enough to withstand earthquakes.

4.3 Interview Survey in 2018

This time, we reflected on the preceding two interview surveys. The main purpose of a series of surveys is to find rules of thumb from people who experienced the 2015 earthquake. However, it was found that if we included earthquakes or preparedness in the question, people would try to come up with only those answers that they think are related to earthquakes or preparedness. In other words, it seemed they try to find answers related to science. Based on this observation, the questions were revisited in 2018. The survey has been conducted in both Chautara and Melamchi in March 2018. The method of the interview was the same as before. We were able to conduct interviews with 11 out of the 18 interviewees of 2016. In addition, we interviewed a family member of four of the interviewees. In total, 15 responses could be collected.

The revised questions were 'What was the worst experience after the earthquake to date? Was this caused by the earthquake?' and 'What is the most important thing not to repeat the worst experience related to the earthquake in the future?'

In response to these questions, most people mentioned the loss of their house as the worst experience. Moreover, some people also mentioned the loss of their family members. However, a man who lost his house mentioned the worst experience was the earthquakes but not the loss of his house. He got scared of the earthquake and aftershocks, but his life was fine as he received a lot of external support. An INGO came to his village, and the residents could choose between receiving either governmental support or INGO support. The governmental support means 300,000 Nepal Rupees in total. Even if it is not sufficient for reconstruction, the government will not give more. INGO support in this village meant the INGO gave all of the construction materials to the people. They also gave them training in how to construct a house. In this case, people could get all of the materials to construct the house, but they had to construct the house by themselves. The picture shows the house provided by the INGO (Fig. 4).

The other man said he had to sell his land to reconstruct his house as the governmental support was not sufficient. However, he also mentioned that even after selling their land, people should construct a strong house in order to avoid a repetition of the worst experience.

Others mentioned different things. A lady whose worst experience was her father and a brother both getting severely injured by the earthquake said proper knowledge is needed. Before the earthquake, many people knew about the drop, cover and hold on. However, she said this is not proper knowledge in Nepal as many people died because they did not escape from their vulnerable houses. She thinks they need a different way of dealing with earthquakes.

Another lady mentioned that some people, including her father, became alcoholics because of the earthquake. She thinks mental support after disasters would be useful to avoid this worse experience.

A man mentioned the importance of experience-based knowledge. He lost his house, and to avoid repeating this experience, he thinks sharing the experience is important. He appreciated scientific knowledge, but he believed experience-based

Fig. 4 The house provided by the INGO

knowledge is the best. He has already started to share his experience with others, in particular the younger generation, in informal settings.

Another man who lost his son and his house thinks that, to avoid a repetition of the worse experience, social harmony is needed. This is because, according to him, if the society is harmonised, then people can come up with new ideas to overcome the problem.

However, the totally opposite idea should also be mentioned. A lady told us she did not get any mutual assistance, and this was one of her worst experiences. Thus, she thinks that not expecting mutual assistance is important to avoid a repetition of the worst experience in the future.

5 Conclusion: Need for Long-Lasting Multilateral Surveys

In order to unfix the relationship regarding disaster education, the possibility of mutual communication in the recovery process of the 2015 Nepal Gorkha Earthquake has been discussed. As an important part of mutual communication, transferring knowledge from those who are not supposed to have scientific knowledge is focused on. In other words, we tried to discover knowledge from the affected local people in Nepal. From a series of interviews, as of now, it is difficult to clearly point out what are the rules of thumb of the local people. However, the following fact should be appreciated. Although they had experienced the disaster and most of them lost their family and/or house, they have started to recover from the disaster. This means the people have somehow overcome the disaster through their efforts. It can be understood that the phrase by a man that is 'experience-based knowledge is useful' is premised on the existence of rules of thumb. It is possible to propose a hypothesis that there have been rules of thumb in the recovery process. It seems the recovery process is slow; some interviewees mentioned the government process took time. However, whether it is a problem or not should be evaluated by the people in Nepal.

In 2016, most people mentioned ways of reconstruction of houses as the preparedness after the earthquake. However, during the interview in 2018, most people mentioned various ideas for avoiding a repetition of their worst experience in the future. These include not only how to build or what kind of house to build but also the importance of localised knowledge, importance of sharing the experience with the younger generation, and the importance of social harmony to generate new ideas. This difference between 2016 and 2018 implies at least two points. The first is people's outlook becomes wider as time passes. More diverse ideas can be seen during the survey in 2018 than in 2016. Second, finding rules of thumb is not easy. The big difference between the survey in 2016 and 2018 is in the wording of the questions. We have used a vaguer expression in 2018 and found a variety of ideas from the survey. As Shiroshita (2018) pointed out, based on the black swan theory, it is not easy to find a thing that people can see but do not realise it. From these two implications, it can be said that long-lasting multilateral surveys should be kept conducting in the affected area to clearly point out the rules of thumb.

If the affected people in Nepal are able to transfer any rules of thumb based on their experience, it will be useful for DRR. This is because the rules of thumb would be able to redeem the limitations of science and science-based technology. Moreover, this fact would encourage the affected people as they will be transformed from those who have not to those who have.

References

Government of Nepal (2018) Nepal Disaster Risk Reduction Portal

http://drrportal.gov.np. Accessed 30 April 2018

http://www.moha.gov.np/uploads/publicationFiles/9740_be8e4c7c9f91b2cf707e43abc4ce6df2_1491503474_N.pdf. Accessed 30 April 2018

https://web.pref.hyogo.lg.jp/kk42/pa20_000000016.html. Accessed 30 April 2018

https://www.unisdr.org/campaign/resilientcities/assets/documents/privatepages/Community-Based%20Earthquake%20Preparedness%20in%20Nepal%20-%20A%20Matter%20of%20Risk%20Perceptions.pdf. Accessed 30 April 2018

Hyogo Prefectural Government (2005) Regarding the survey on casualties of the Great Hanshin-Awaji Earthquake Disaster

Intensive Study and Research Center (2013) District and VDC profiles of Nepal—2013. Intensive Study and Research Center, Kathmandu

International Federation of Red Cross and Red Crescent Societies (2014) World disasters report 2014—focus on culture and risk. http://www.ifrc.org/Global/Documents/Secretariat/201410/WDR%202014.pdf. Accessed 30 April 2018

Laursen MR (2015) Community-Based earthquake preparedness in Nepal: a matter of risk perceptions

Ministry of Culture, Tourism & Civil Aviation (2016) Nepal tourism statistics 2016. http://www.tourism.gov.np/downloadfile/Nepal%20Tourism%20statistic_Final-2016_1498990228.pdf. Accessed 30 April 2018

Ministry of Finance (2018) Post-earthquake assistance portal. http://earthquake.mof.gov.np/portal/. Accessed 30 April 2018

Ministry of Home Affairs (2015) Nepal earthquake 2072: situation update as of 11th May. http://drrportal.gov.np/uploads/document/14.pdf. Accessed 30 April 2018

Ministry of Home Affairs (2016) Disaster risk reduction in Nepal: achievements, challenges and ways forward

Oven K, Milledge D, Densmore A, Jones H, Sargeant S, Datta A (2016) Earthquake science in DRR policy and practice in Nepal. Overseas Development Institute. https://www.odi.org/sites/odi.org.uk/files/resource-documents/10638.pdf. Accessed 30 April 2018

Preston J (2012) Disaster education: 'Race', equity and pedagogy. Sense Publishers, Rotterdam

Sato H (2005) Sociology for development assistance [Kaihatsu Enjo no Syakaigaku]. Sekaishisosha, Kyoto

Shiroshita H (2012) Safety education in the UK: multilayer education in safety centres. J Jpn Soc Civ Eng Ser. F6 (Safety Problem) 68(2):146–152

Shiroshita H (2014) Towards the actualisation of co-learning for disaster prevention [bosaikyoiku no jitsugen ni mukete]. In Faculty of Safety Science, Kansai University (Ed.), Safety Science for disaster prevention and reduction—a proposal in building a safe and secure society [bosai/gensai no tameno shakaianzengaku—anzen/anshin na shakai no kochiku eno teigen] Mineruva Shobo, Kyoto, pp 98–114

Shiroshita H (2018) Is disaster education just knowledge transmission? co-learning and disaster governance in Japan. In Forino G, Bonati S, Calandra LM (eds) Governance of risk, hazards and disasters trends in theory and practice, Routledge, London, pp 274–277

The World Bank Data (2016) GDP per capita (current US$). https://data.worldbank.org/indicator/NY.GDP.PCAP.CD?year_high_desc=true. Accessed 30 April 2018

Thiruppugazh V (2015) Disaster management capacity building. http://www.icnr2015.mof.gov.np/uploaded//Disaster_Management_Capacity_Building_24th_June.pdf. Accessed 10 April 2016

Tuladhar G, Yatabe R, Dahal R, Bhandary N (2015) Disaster risk reduction knowledge of local people in Nepal. Geoenvironmental Disasters. https://doi.org/10.1186/s40677-014-0011-4

UNESCO (2012) Education for disaster preparedness. http://www.unesco.org/archives/multimedia/?pg=33&s=films_details&id=2795. Accessed 30 April 2018

UNISDR (2012) What is disaster risk reduction? https://www.unisdr.org/who-we-are/what-is-drr. Accessed 30 April 2018

United Nations (2015) Sendai framework for disaster risk reduction 2015–2030. https://www.preventionweb.net/files/43291_sendaiframeworkfordrren.pdf. Accessed 30 April 2018

Yamori K, Onishi M (2016) <Disaster 1.0> and <disaster 2.0> [<Saigai 1.0> to <Saigai 2.0>]. In: Proceedings of the 35th annual conference of japan society for natural disaster science, pp 61–62

Imminent Warning Communication: Earthquake Early Warning and Short-Term Forecasting in Japan and the US

James D. Goltz and Evelyn Roeloffs

Abstract The ability of scientists to warn communities of the occurrence of potentially damaging earthquakes is evolving, though accurate prediction of the time, location, and magnitude of such earthquakes remains a distant goal. Anticipation of major earthquakes has taken the form of long-term forecasts, earthquake hazard mapping, and scenario development accompanied by societal impact assessments based on loss estimation methodologies. More recently, earthquake early warning systems (EEW), pioneered and first implemented in Japan, facilitated the communication of very short-term alerts that ground motion from an earthquake would arrive in a matter of seconds to tens of seconds. Currently, a nationwide EEW system exists only in Japan, but regional systems are becoming available in many earthquake vulnerable nations, including the US. A second means of anticipating an earthquake is operational earthquake forecasting (OEF) based on the occurrence of seismic activity of various kinds that raises the short-term probability that additional earthquakes, including damaging earthquakes, could occur in hours to a few days. OEF has been implemented in the US and has its most vocal champions among American scientists. These two means of anticipating potentially damaging earthquakes and the communication of information to communities at risk are the subject of this chapter.

Keywords Earthquakes · Early warning · Short-term forecasts · Japan · United States · Warning

1 Introduction

Scientists, emergency managers, and communities at risk of damaging earthquakes have long hoped for a breakthrough that would provide accurate predictions of the time, location, and magnitude of future earthquakes. Regrettably, we are unlikely to

---

J. D. Goltz (✉)
Disaster Prevention Research Institute, Kyoto University, Kyoto, Japan
e-mail: jamesgoltz@gmail.com

E. Roeloffs
United States Geological Survey, Earthquake Science Center, Vancouver, Wa, USA
e-mail: evelynr@usgs.gov

K. Yamori (ed.), *Disaster Risk Communication*, Integrated Disaster Risk Management,
https://doi.org/10.1007/978-981-13-2318-8_8

realize this breakthrough any time soon. But we currently have two strategies for alerting communities that may experience ground motion from an earthquake in an imminent time frame, earthquake early warning (EEW), and short-term operational earthquake forecasting (OEF). In this chapter, these two technologies will be discussed as they have been, or will be, implemented in the United States and Japan. Our emphasis will be on risk communication as facilitated by these technologies rather than the technical details of the technologies themselves.

At a basic level, an EEW system provides notification that an earthquake is occurring and that potentially damaging ground motion is approaching a particular region or community. The US West Coast "ShakeAlert" system and the nationwide EEW system in Japan (Kinkyu Jishin Sokuho) each rely on a network of seismic stations continuously transmitting data to a processing site where successively refined estimates of an earthquake's magnitude and location are computed in near real time as the seismic rupture progresses. If the expected ground motion from the earthquake exceeds a threshold value, an alert is transmitted from the data processing site to users in the area at risk. Warning times depend on the distance between the earthquake source and the populated area and may vary from no warning at all to more than a minute if the source is quite distant (Allen et al. 2009).

Short-term operational earthquake forecasts (up to several days) are issued based on the occurrence of regional seismic activity that has increased the probability of large damaging level events in nearby locations. Short-term operational earthquake forecasting is based on the following scientific observations: earthquakes tend to cluster in time and space (e.g., aftershocks are triggered by large earthquakes); earthquake probabilities are not constant in time, but rather rise and fall over periods of days to years in correlation with nearby seismic activity; as a sequence of aftershocks following an initial earthquake unfolds, sometimes a larger earthquake occurs (e.g., the Kyushu, Japan sequence in April 2016); and although such foreshock–mainshock sequences are rare, depending on the region, short-term probabilities for larger earthquakes in 3–5 days may range up to 10%.

Short-term operational earthquake forecasting and EEW are independent technological approaches, but could potentially supplement one another in an earthquake emergency. EEW depends on a dense seismic network, real-time telemetry, highly efficient algorithms, and rapid communication channels to deliver a message in seconds, warning that ground motion from an earthquake is approaching a community or an EEW user location in seconds to tens of seconds; this EEW infrastructure must be built to function during and after a major earthquake. Short-term operational earthquake forecasting, based on statistics of past seismicity, can only yield low daily probabilities of damaging events, as only a small number of earthquakes later prove to be foreshocks of destructive earthquakes (Reasenberg and Jones 1989, 1994; Reasenberg 1999). One can speculate, however, that in regions where both technologies are in place, receiving a forecast that there is a slight short-term (up to several days) increase in earthquake potential would heighten awareness of the imminent possibility of an earthquake and, if a potentially damaging event were to occur, people might have planned how to use seconds of early warning more effectively.

In the sections that follow, we trace the legal and historical development of these technologies in Japan and the US, explore their implementation including lessons learned and changes made and examine the risk communication implications of these earthquake alerting strategies as they exist, or in the case of EEW in the US, as they are currently being prepared for public availability.

2 Earthquake Early Warning: Japan and the US West Coast

2.1 Earthquake Early Warning in Japan

Although Japan has employed EEW systems in varying capacities, most notably on the Shinkansen, for approximately 50 years, the effort to develop a nationwide comprehensive system began in 2003 and culminated in implementation of the current system (Kinkyu Jishin Sokuho) in October 2007. As such, it became the first nationally available EEW system in the world. The EEW system was a cooperative effort of several Japanese government agencies including the Ministry of Education, Culture, Sports, Science and Technology (MEXT), the National Research Institute for Earth Science and Disaster (NIED), the Japan Meteorological Agency (JMA) and nongovernmental organizations including the Real-Time Earthquake Information Consortium (REIC) and the Japan Weather Association (JWA). Following the 1995 Hanshin-Awaji (Kobe) earthquake, the essential infrastructure for the EEW system, the high-sensitivity seismographic network (Hi-Net) of NIED, and the seismic monitoring system of JMA were developed and became operational. Although NIED and the JMA operate other seismic networks in Japan, only the Hi-Net system and the JMA's Tsunami and Earthquake Observation Monitoring networks transmit data in real time, essential for EEW (see Fig. 1). The two networks include approximately 1000 stations spaced at approximately 25 km intervals (NIED 2012; Fujinawa and Noda 2013). Warnings are issued to the public when the maximum estimated intensity is expected to reach 5-Lower on the JMA intensity scale (Allen et al. 2009; Fig. 2).

In 2006, prior to the full implementation of the EEW system, the JMA initiated a public information campaign to acquaint potential EEW users with the system's capabilities and limitations. This information campaign included brief informational videos shown on television, brochures, posters, and web site information; please see http://www.jma.go.jp/jma/en/Activities/eew.html. These information materials are oriented to various users but have a common theme in inducing those who receive warnings to take emergency action prior to the arrival of strong ground motion at home, school, or work such as seeking cover under sturdy furniture and moving away from potential hazards. EEW users are also advised that warnings may be employed to slow or stop trains to avoid derailments, control elevators to prevent entrapment, prevent traffic accidents by advising drivers to pull to the side of the

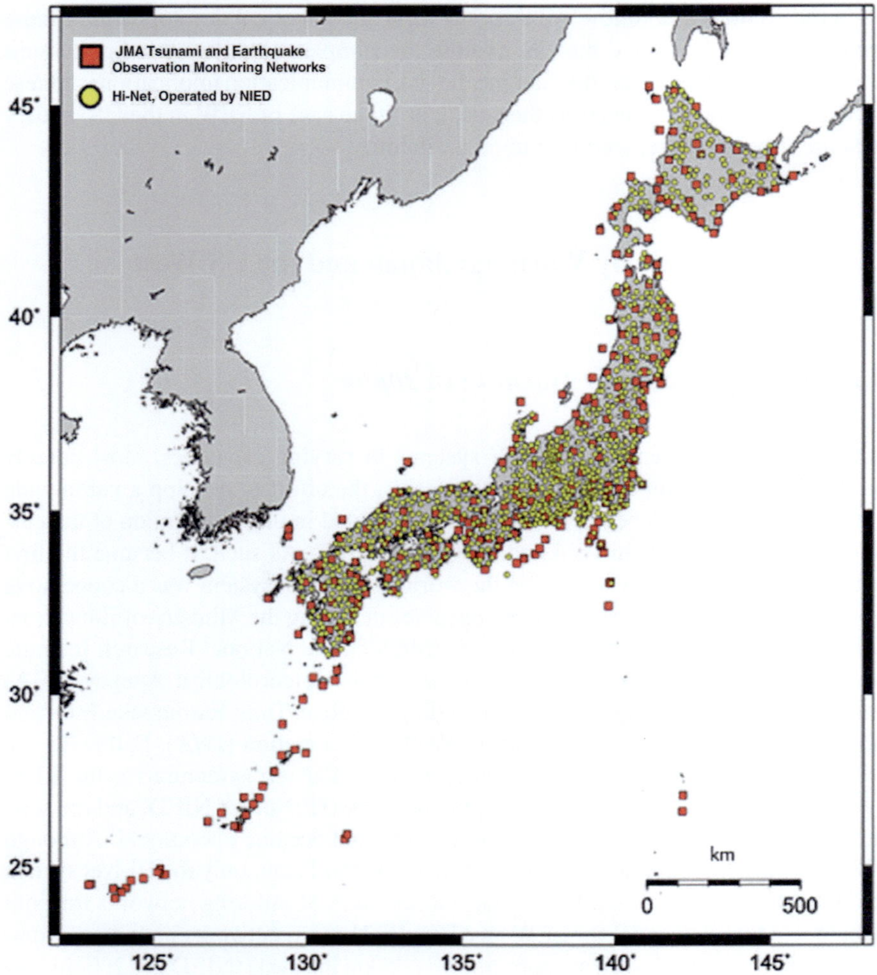

Fig. 1 Seismic networks for earthquake early warning in Japan. Hi-Net, operated by NIED (yellow circles), and JMA Tsunami and Earthquake observation monitoring networks (red squares) that transmit data in real time. *Source* Japan Meteorological Agency (https://www.data.jma.go.jp/svd/eew/data/nc/shikumi/shikumi.html). Modified: original label in upper left translated to English from Japanese

road and turn on hazard lights and suspend work in progress to avoid errors (surgery presented as an example). Potential EEW users are also advised of the system's shortcomings including the very short duration of warnings (several seconds to a few tens of seconds), the fact that those close to the earthquake source may not receive any warning and the potential for errors—under or overestimation of the shaking intensity, false alarms, or missed events (JMA Observed 2017).

JMA Intensity	0	1	2	3	4	5 Lower	5 Upper	6 Lower	6 Upper	7
MMI Intensity	1	2	3	4	5	6	7	8	9	9

Fig. 2 Comparison of the JMA seismic intensity levels with Modified Mercalli intensities. Reproduced from Musson and Cecic (2012) with permission of the authors. (This table is not based on empirical values but on comparison of the wording of the written versions of the scales. In the case of the MMI scale, this is neither the 1931 version (Wood and Neumann 1931) nor the 1956 version (Richter 1958), but the more recent Stover and Coffman version, which was not actually published by Stover and Coffman (1993), but was reconstructed and published in Musson and Cecić (2012), and is the version to which DYFI relates. Recent intensity values are derived from instrumental ground motion recordings (JMA) or Internet-based algorithms (MMI) and not the written scales.)

The JMA EEW messages are communicated to various groups through multiple channels. All NHK (Japan Broadcasting Company) affiliated television and radio stations are required to transmit warnings and nearly all other television and radio stations follow suit. Launched in 2007, J-Alert is a multi-hazard satellite-based system designed to quickly broadcast alerts to municipal governments which in turn relay messages to local media and residents via loudspeakers, television, radio, and cell phones. In 2009, the system was expanded beyond municipal governments to include national administrative agencies (e.g., JMA), regional administrative agencies (e.g., regional police bureaus), courts and designated private sector organizations. The number of municipal governments covered by J-Alert expanded significantly after the March 2011 Great East Japan Earthquake and Tsunami from approximately 46% in 2011 to 93.5% by 2014 or 1,629 of 1,741 jurisdictions (Japan Ministry of Internal Affairs and Communications 2014). Japan's three major mobile phone carriers: NTT DOCOMO, au (KDDI and Okinawa Cellular), and Softbank Mobile have developed simultaneous broadcast systems that allow millions of cell phone users to receive SMS text messages with EEW information.

The JMA offers two types of EEW messages, one (EEWg) for use by the general public and one for private service providers certified by the JMA to relay messages to advanced users (EEWa) (Fujinawa and Noda 2013; see Fig. 3). The JMA divides Japan into 188 seismic hazard blocks, an average of four blocks per prefecture. An EEWg is announced in all blocks where the strength of shaking is judged to be greater than 4 on the JMA intensity scale though at least some blocks must be expected to exceed 5-Lower. The informational content of the EEWg message includes the time of occurrence, epicentral location, names of the prefectures where stronger shaking is predicted and verbal and written warning messages ("please prepare for powerful tremors"). In addition to the criterion of expected shaking of JMA 5-Lower, a second criterion for issuing EEWg alerts is that seismic activity be detected at 2 or more seismic stations. EEWg alerts are also available for the hearing impaired (op cit., p. S347).

EEWa is for advanced users of earthquake early warning. The JMA does not provide site-specific warning information but once certified by the JMA, private

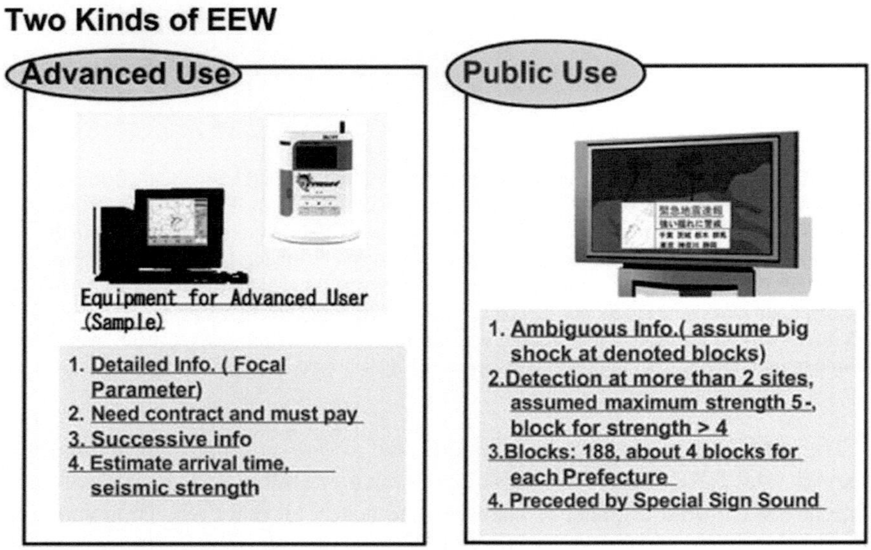

Fig. 3 Basic features of EEWg and EEWa with samples of the specific terminals for EEWa and the graphic display for EEWg. Reproduced from "Japan's Earthquake Early Warning System on 11 March 2011: Performance, Shortcomings, and Changes" by Fujinawa and Noda (2013), Earthquake Spectra, Volume 29, Issue S1, with permission of the Earthquake Engineering Research Institute

service providers can transmit this information mainly to organizations that include rail companies, factories, construction companies, residential sites, hospitals, and educational institutions. An EEWa is issued when the shaking intensity is estimated to be greater than 3 or when the magnitude is greater than Mj3.5. The private EEW service providers offer a variety of services that range from a simple translation of the JMA data into site-specific estimated intensity to sophisticated systems that utilize EEW parameters to modify or curtail processes that would lead to injuries and significant economic losses if continued during strong shaking (Allen et al. 2009). The Japan Meteorological Business Supporting Center is one EEW service provider and receives a fee for its services from EEWa users.

Prior to its public release in Japan, the JMA tested the EEW system, logging nearly 2,000 activations during which 30 errors were recorded. The errors were attributed to lightning, site noise, detection algorithm problems, and instrument failures. During this testing period, latency periods (the length of time between event detection and issuance of a warning) ranged from 1 to 11 s; the mean elapsed time for one-station data was 5.4 s, compared with 6.4 s for multi-station data. Accuracy of predicted ground motion was another evaluation criterion; 45% of activations showed agreement between predicted and actual shaking intensity and 91% were accurate within ±1 on the JMA shaking intensity scale (Fujinawa and Noda 2013). During its first 2 years of operation (2007–2009), warnings were issued for 11 events in which JMA intensity 5-Lower or 5-Upper was observed. Of these events, two missed events were recorded in which the maximum predicted intensity was 4 but 5-Lower was observed, and three false alarms in which 5-Lower was predicted but only intensity 4 was reached (Allen et al. 2009).

The Mj9.1 Great East Japan (Tohoku) Earthquake and Tsunami of March 11, 2011

Since the EEW system became operational in October 2007, the most significant earthquakes for which warnings were issued were the Mj9.1 Great East Japan (Tohoku) Earthquake and Tsunami on March 11, 2011 and the series of earthquakes in Kyushu, culminating in the JMA Mj7.3 Kumamoto earthquake of April 16, 2016. In one of the most significant earthquakes of the last century, the Mj9.1 Great East Japan Earthquake and Tsunami claimed the lives of approximately 18,000 people. The Japan Meteorological Agency issued the first EEW at 14:46:45.6, 5.4 s after detection for a Mj4.3 event, which exceeded the threshold for advanced users (Mj3.5); however, due to the low estimated intensity (JMA 1) no user actions were triggered. The second EEW was issued 1.1 s after the first for a Mj5.9 earthquake, but no public warning was yet issued because the predicted seismic intensity did not exceed JMA 5-Lower. The first public warning was the fourth EEW, for a Mj7.2 event and JMA 5-Lower shaking intensity, issued for central Miyagi Prefecture at 14:46:48.8, 8.6 s after detection. Subsequently, 11 additional updated EEWs were issued, culminating in a projected JMA magnitude of 8.1 and 6-Lower shaking intensity at 14:48:37.0, 116.8 s after detection. People in the most heavily shaken areas (Miyagi, Iwate, and Fukushima Prefectures) would have received at least 10 s of warning prior to the arrival of strong ground motion. The Kanto area (including Tokyo) also experienced strong shaking but the region did not receive EEW alerts, leading many to question the usefulness of the early warning system (Fujinawa and Noda 2013).

The vast majority of fatalities in the Tohoku disaster were not caused by earthquake ground motion, but by the massive tsunami that followed, generated by the earthquake. The JMA issued a tsunami warning at 14:49, approximately 3 min after the earthquake was detected. This warning was for a tsunami of 6 m in Miyagi, 3 m in Iwate and Fukushima and 1 m on the Pacific Coast of Aomori prefectures. These estimated wave heights were derived from computer simulations of tsunamis for scenario earthquakes of various locations and magnitudes. The JMA was unable to update the tsunami warning with a moment magnitude from broadband seismometers, because these instruments saturated. However, a GPS buoy deployed in the ocean 10 km offshore from Kamaishi City (Iwate Prefecture) recorded a rapid increase in sea level approximately 24 min after the earthquake, and on this basis, the JMA issued a second tsunami warning at 15:14 projecting wave heights of over 10 ms for Miyagi, 6 m for Iwate and Fukushima, and 3 m for the Pacific Coast of Aomori Prefectures. A third tsunami warning, based on GPS buoys and tide gauges was issued at 15:30 expanding the areas warned with projections of over 10 m surges for Miyagi, Iwate, and Fukushima and 8 m for parts of Aomori (JMA 2013).

The April 2016 Kumamoto earthquake sequence

A vigorous earthquake sequence began on the island of Kyushu, Japan in the vicinity of Kumamoto on April 14, 2016. An initial earthquake occurred at 21:26 on April 14 with a JMA magnitude of 6.5, followed by many aftershocks including one at Mj6.4.

On April 16 at 01:25, a Mj7.3 event occurred near the epicenter of the first earthquake causing JMA intensity 7 ground shaking. The JMA reported that between April 14 and April 30, the number of felt earthquakes in Kumamoto and Oita Prefectures reached 1093 (JMA 2016). The main events in the sequence caused 49 fatalities and caused extensive damage in Kumamoto City and a number of smaller towns in Kumamoto and Oita Prefectures. Between April 14 and 19, JMA issued 19 EEWs during the height of the earthquake sequence. For the Mj6.5, the system issued an initial EEW 3.8 s after detection, location error was less than 10 km, and the estimated magnitude was equal to the actual magnitude. All areas where the maximum observed intensities exceeded JMA 4 were covered in that EEW, which predicted a maximum intensity of 6-Upper (Kodera et al. 2016). An initial EEW for the Mj7.3 mainshock was issued 3.9 s after detection with an epicentral error of less than 10 km and a magnitude of Mj5.9.

Kodera notes that although the magnitude was underestimated, it was large enough to trigger a warning (p. 4). A second EEW was issued 4.7 s after the first, with an estimated magnitude of Mj6.9. Both EEWs covered all areas with seismic intensities exceeding JMA 5 (op cit., page 4). Among the 19 EEWs issued between April 14 and 19, there were four errors attributed to the occurrence of multiple simultaneous earthquakes, errors in estimated epicentral location, and miscalculation of seismic intensity distribution. In addition, there were four earthquakes between April 14 and 30 that generated ground motions exceeding 5-Lower for which EEWs were not issued. The system did detect these events, but the predicted maximum intensities did not reach 5-Lower.

Improvements to Japan's EEW system

The 2011 Tohoku earthquake and tsunami and the 2016 Kumamoto earthquake sequence, as well as less significant earthquakes that have occurred since the implementation of Japan's EEW system, have resulted in a number of lessons learned and changes designed to improve system performance. The EEW system problems recognized following the Great East Japan (Tohoku) Earthquake and Tsunami include: an underestimation of the size and impact of the Mj9.1 earthquake and its ensuing tsunami, delays in calculating an M_w due to broadband seismic data saturation (JMA 2013), errors caused by the simultaneous occurrence of multiple earthquakes (Fujinawa and Noda 2013), and delays due to equipment damage and electrical outages. In addition, TV and radio stations experienced delays of several minutes in broadcasting EEW messages to users.

The JMA implemented several changes after the 2011 Tohoku earthquake. They installed 80 broadband strong-motion seismometers to record seismic waveforms without saturation even in very large earthquakes. In 2011, a large-scale project was launched to install the Seafloor Observation Network for Earthquakes and Tsunamis along the Japan Trench, consisting of 150 ocean bottom observation stations. Seafloor observations from this network, completed in 2015, will reduce delays in tsunami warnings and enable more precise wave height prediction (Uehira et al. 2015). In addition, three DART buoys were installed off the coast of Tohoku, which in conjunction with ocean bottom pressure sensors deployed by related organizations, will provide more rapid detection of tsunamis, facilitate estimation of tsunami wave heights and improve tsunami warning capabilities. To avoid data loss from damage to seismic network equipment and power outages, the JMA installed satellite-linked telecommunications equipment as a backup to terrestrial communications (JMA 2013). Following the Kumamoto earthquake sequence, the issue of under- and overprediction errors was addressed by the implementation of integrated particle filter (IPF) and propagation of local undamped motion (PLUM) methods to address overprediction due to the simultaneous occurrence of multiple earthquakes and underprediction for earthquakes with large rupture zones (Kodera et al. 2016). The JMA added an EEW category "major emergency", for earthquakes expected to be JMA intensity 6-Lower or 6-Upper and tsunami heights projected to exceed 3 m (JMA Observed 2017).

In addition to improvements to the EEW system, there were significant changes in public perception and salience of the system as revealed in surveys conducted before and after the Great East Japan (Tohoku) Earthquake and Tsunami (JMA 2012; Ohara et al. 2011). A Tokyo University-based survey (Ohara et al. 2011) was conducted with residents of metropolitan Tokyo, including Kanagawa, Saitama, and Chiba Prefecture residents 30–60 years of age both before (November 2010) and after (August 2011) the Tohoku earthquake. The nonrandom sample included, in both surveys, approximately 700 adults who responded to an Internet questionnaire. A separate survey, conducted by the JMA approximately a year after the Tohoku earthquake, employed both Internet and postal mail questionnaires to select a final (nonrandom) sample of 1,600 EEWg users and 400 EEWa users. The samples were drawn from the 22 prefectures whose residents had received EEW messages during the Tohoku earthquake (JMA 2012). In this survey, the sample was divided into EEWg users, EEWa users, and a third subsample of 817 residents in the Tohoku region (Iwate, Miyagi, and Fukushima) who experienced the most severe shaking (JMA 7). The Tokyo University and JMA survey questions were similar, covering EEW awareness, knowledge of operation and limitations, actions taken in response to the warnings received, whether respondents had planned any specific action prior to receipt of the warnings, and overall evaluation of the usefulness of EEW.

The pre-Tohoku survey conducted by Ohara and colleagues at Tokyo University revealed that just 50% of those surveyed were aware of the EEW system, despite considerable efforts by the JMA to disseminate information about it, and promote its use. That figure increased dramatically to 87% in both post-Tohoku surveys. An equally large majority of post-Tohoku survey respondents understood the basic function of the EEW system for earthquakes (that strong shaking can be expected in a very short time when EEWs are issued) compared to just 34% prior to the March 2011 disaster (Ohara et al. 2011). The increased awareness was attributed to the experience of the Tohoku earthquake and over 400 significant aftershocks. Both post-Tohoku surveys revealed that only a small percentage of respondents (~20%) understood the difference between the early warnings designed for advanced users compared to warnings for the general public. Low salience of this difference was attributed to less experience with EEWa and the cost involved in use of EEWa.

The JMA survey (2012) revealed that a great majority of survey respondents (78%) attempted to take action based on receipt of an EEW alert; however, only a subset of respondents (61%) reported success in taking the action intended, while 17% reported being unsuccessful in taking the intended action. Those who took no action were those who had no knowledge of the EEW system or regarded it negatively. This survey asked respondents which of the following actions they took in response to the warnings received: took refuge under a desk or furniture to avoid hazards; protected children and older persons; opened doors and went outside; obtained information through TV and/or radio; and shut off gas or heater. The most frequent actions taken were to seek protection for oneself and protect others, selected by a majority of survey respondents in each of the three subgroups (EEWg, EEWa, and strongest shaking). The subsample of respondents who had experienced the strongest shaking was twice as likely as the other subsamples to indicate that they also opened doors and went outside and shut off gas and heaters. Those who attempted to learn more about the situation through radio or television ranged from 22% among general users to 26% for the strongest-shaking subsample. The JMA survey also revealed considerable satisfaction with the early warning system: 82% of EEWg users, 91% of EEWa users, and 90% of the strongest-shaking group rated it as "useful" or "somewhat useful". The nature of this assessed usefulness was in the "ability to stand ready" for large majorities of each subsample. In contrast to the other subsamples, the strongest-shaking group rated the "ability to protect themselves" as a second benefit while the EEWg and EEWa groups selected "psychological merit" as the more significant secondary benefit (Ohara et al. 2011).

The experiences of the Great East Japan Earthquake and Tsunami of 2011 and the Kumamoto sequence of 2016 had significant impacts on Japan's EEW system prompting improvements in the system's accuracy and reliability and, based on survey results, increasing the salience of the EEW system among Japan's diverse EEW users. These changes are likely to improve both use of and confidence in the EEW system and are likely to be adopted by the ShakeAlert system developers on the West Coast of the United States and those anticipating the implementation of EEW systems around the world.

2.2 Earthquake Early Warning in the United States (West Coast)

The first known mention of EEW in the United States appeared as a "letter to the editor" in the San Francisco Daily Chronicle from a San Francisco Bay area medical doctor in 1868. The device envisioned by Dr. J.D. Cooper would use electrical current to ring an earthquake bell in the city alerting residents that seismic waves were approaching, allowing them time to escape dangerous buildings (Allen et al. 2009). Apparently, no one was ready to implement such a device at the time and the next discussion of EEW appeared in the 1980s, concurrent with the prospect of scientific earthquake prediction. In 1985, an article appeared in the journal *Science* advocating the development of a "Seismic Computerized Alert Network" that would "provide short-term warning (as much as several tens of seconds) of imminent strong ground motion from large earthquakes" (Heaton et al. 1985). In 1988, a survey was conducted by the California Division of Mines and Geology (now the California Geological Survey) in which various organizations were asked whether a system that provided a few seconds of advanced warning prior to the arrival of earthquake ground motion could be beneficial in reducing damage to corporate facilities and production, thus saving large sums of money (Holden et al. 1989). The conclusion was that it would not, though the possibility that such a system could save lives, if not dollars, was not explored by the survey.

The first effort to actually utilize EEW in the U.S. followed the 1989 M_w6.9 Loma Prieta earthquake, which occurred 100 km south of the cities of San Francisco and Oakland. Both cities experienced building and roadway collapses owing to unfavorable site conditions. Because search-and-rescue and repair crews were at risk from potential shaking due to strong aftershocks, and because the source area of the shaking was located some distance from the workers at risk, the USGS built and deployed a prototype EEW system to alert these workers of incoming strong shaking. This temporary prototype system worked successfully, but other than documentation of the effort in the Bulletin of the Seismological Society of America (Bakun et al. 1994), no efforts were initiated to build a permanent EEW system. Following the 1994 Northridge earthquake, the TriNet Project was initiated by the California Institute of Technology, the United States Geological Survey and the California Division of Mines and Geology, designed to enhance and upgrade the Southern California Seismic Network. Included in the TriNet proposal to the Federal Emergency Management Agency was "a pilot earthquake early warning project in southern California", a component of the proposal that was later abandoned because the hardware and software were not available at the time (Goltz 2003). It was not until 2006 that the USGS began to fund research and development toward the goal of an EEW system for the state of California.

With USGS funding and additional funds from the state of California, scientists and engineers at the California Institute of Technology (Caltech), University of California (UC) Berkeley, the Swiss Federal Institute of Technology (ETH), and the University of Southern California (USC) began development of an EEW demonstration system for California, called CISN (California Integrated Seismic Network) ShakeAlert (Böse et al. 2014; see Fig. 4). This demonstration system utilized real-time data from about 380 broadband and strong-motion stations throughout California. In 2009–2011, funds from the American Recovery and Reinvestment Act were used to upgrade equipment to allow seismic waveform data to reach three processing facilities (Caltech/USGS Pasadena, UC Berkeley, and USGS Menlo Park) in seconds. Development of the US West Coast EEW system has proceeded although funding for a fully operational system is not yet in place. In 2011, it was estimated that an EEW system for California would cost $80 million over 5 years, and including Oregon and Washington would add $70 million over that time period. In late 2011, the Gordon and Betty Moore Foundation provided $6 million in grants to the University of California Berkeley (UCB), the California Institute of Technology (Caltech), the University of Washington (UW), and the U.S. Geological Survey

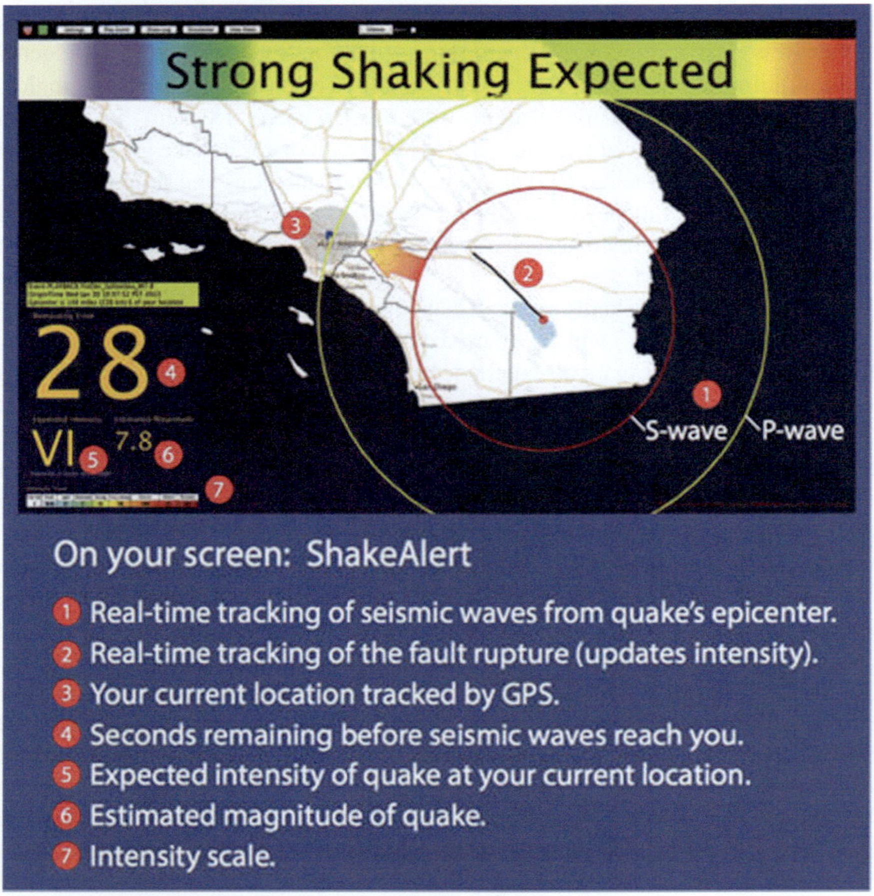

Fig. 4 The features of ShakeAlert in which a user receives a message like this on the screen of his or her computer. The message alerts the user to how many seconds remain before the shaking begins at their location and the expected intensity of shaking at that site. The shaking intensity follows the Modified Mercalli scale; an intensity of VI, as shown here, would mean that the shaking is felt by everyone, people find it difficult to stand, and structures may suffer some damage. The warning message also displays a map with the location of the epicenter, the current estimate of the magnitude of the quake, and the current position of the P- and S-waves. In this example, the alert is for a hypothetical scenario of a magnitude 7.8 earthquake on the southern San Andreas Fault. Courtesy of the United States Geological Survey, U.S. Department of the Interior and cited from https://earthquake.usgs.gov/research/earlywarning/overview.php. Accessed 17 Jan 2019

(USGS) to develop and coordinate a West Coast-wide prototype EEW system based on the CISN ShakeAlert system (Hartog et al. 2016).

The CISN ShakeAlert demonstration project targeted beta testers from the scientific community as well as users who could implement automatic responses; it was not funded to develop the educational materials needed to enable the public to utilize EEW information. Target users included the California Office of Emergency Services (CalOES) and the Bay Area Rapid Transit District (BART) in the San Francisco Bay Area. CISN ShakeAlert began sending warnings to these test users in January 2012, and as of September 2012, BART could use CISN ShakeAlert warnings to slow trains to speeds that reduce the potential for derailment during seismic shaking (BART News Release 2017). Other test users include local emergency management agencies, hospitals, various utilities, entertainment venues, universities and school districts, transportation agencies and businesses. Among the criteria for becoming a ShakeAlert test user are having an interest in partnering with the ShakeAlert project to demonstrate the usefulness of the ShakeAlert system; committing sufficient resources to successfully install and run the ShakeAlert UserDisplay application and monitor results; participating in training to understand the capabilities and limitations of the ShakeAlert system; providing suggestions and feedback regarding the ShakeAlert system, including functionality, design, usability, bugs, and results of any testing performed by the user; and understanding that no actions should be taken based on the ShakeAlert notifications (USGS 2016).

On August 24, 2014, the magnitude 6.0 South Napa earthquake struck approximately 50 km northeast of San Francisco, reaching MMI VIII in the epicentral area and MMI IV in the large cities of the San Francisco Bay area. The first CISN ShakeAlert alarm was generated 5.1 s after the event origin, with a magnitude of 5.7, which evolved through successive alerts to M6.0, in agreement with final seismological analyses (Allen et al. 2014). Beta users received the alert, which gave San Francisco 911 operators eight seconds of warning to prepare for the high volume of calls that would be arriving. With 10 s warning at UC Berkeley, campus police relayed a warning via their radio system before the shaking began. Examining how the ShakeAlert algorithms responded to the South Napa earthquake revealed ways to improve performance and demonstrated the value of including GPS data to obtain accurate alerts for events larger than M6.0 (Grapinthin et al. 2014).

The University of Oregon became a partner in development of the West Coast-wide system as part of a Congressional appropriation in late 2015 and partners added since then include the University of Nevada, Reno, Central Washington University and UNAVCO, Inc. (Kohler et al. 2017). The Pacific Northwest (PNW) ShakeAlert system, covering Oregon and Washington, was initially developed as a separate system from CISN ShakeAlert, but these systems were effectively combined in early 2017. A particular challenge for EEW in northernmost California, Oregon, and Washington is to provide accurate detection and magnitude estimation for events originating offshore in the Cascadia Subduction Zone. There are no real-time seafloor seismometers off the coasts of northern California, Oregon, or Washington. The EEW algorithm being tested requires the P-wave to be recorded at four seismic stations, so the delay in alert time depends on the density of the network, which is sparser in Oregon and Washington than near population centers in California. Data acquisition equipment is being upgraded to move data from stations to processing centers more quickly, and new stations are being installed to increase network density (Hartog et al. 2016). To provide robust EEW for the entire US West Coast, it is estimated that 1675 seismic stations are needed at an average spacing of 20 km or less (Given et al. 2014).

Currently, about 670 stations are contributing data to ShakeAlert, but some of these stations require upgrading to prevent clipping during large ground motions (Kohler et al. 2017). In March 2017, the "Pacific Northwest Earthquake Preparedness Act of 2017", a bill requiring funding for and construction of an EEW system for Cascadia, passed the U.S. House of Representatives; as of this writing the bill awaits action by the U.S. Senate.

A select group of test users in Oregon and Washington began to obtain alerts in 2015 (Hartog et al. 2016). In April 2017, an engineering firm that designs municipal water and sewage plants became the first commercial organization in Washington state to sign up to receive EEW alerts, which they will use to automatically close valves to prevent draining of storage tanks should seismic shaking damage water mains (Seattle Times 2017). Oregon's first test user is the Eugene, Oregon Water and Electricity Board, which will use alerts to stop turbines at a river power plant and lower water levels in a canal above a residential area (University of Washington News 2017).

The ShakeAlert Joint Committee for Communication, Education, and Outreach (JCCEO) was formed in 2016 to provide advice regarding the ShakeAlert system's human interface, which is currently under design. The committee will coordinate the development of guidance and resources to maximize the utility of ShakeAlert for various classes of users. In doing so, JCCEO will make recommendations regarding best practices based on user experience and on social science research. JCCEO is managed by a USGS scientist who reports to the National ShakeAlert Program Coordinator and operates with the advice of a national committee with representative personnel from Washington, Oregon, California and British Columbia, Canada. Thresholds for broadcasting alerts are among the issues that remain to be decided.

On January 2, 2019, Los Angeles Mayor Eric Garcetti announced the availability of ShakeAlertLA, an earthquake early warning application for smartphones that is available to residents of Los Angeles County who wish to download the free application. The application sends alerts to users within Los Angeles County indicating that an earthquake of magnitude 5.0 or greater or level IV intensity has been detected and that they may soon feel shaking. The "app" is available for both Apple and Android devices in both English and Spanish. This was the first "roll-out" of ShakeAlert for the general public in the United States. At the end of the first month of operation, this cooperative effort between the City of Los Angeles and the United States Geological Survey resulted in approximately 410,000 downloads. On July 4, 2019 a M_w6.4 earthquake occurred near the City of Ridgecrest in (Kern County) southern California, located approximately 150 miles NNE of Los Angeles. This earthquake was followed by a larger event (M_w7.1) in the same general location on July 5th. Both events were widely felt in Los Angeles County, however, ShakeAlert provided no public warning of approaching ground motion because most of Los Angeles County fell below the intensity threshold of Modified Mercalli intensity IV. In response to public disappointment in not being warned, the City of Los Angeles lowered the threshold to intensity III. On October 17, 2019, the anniversary of the 1989 Loma Prieta earthquake, California Governor Gavin Newsom announced that the United States Geological Survey and its state and university partners were now making ShakeAlert available to the entire state of California. The states of Oregon and Washington will be included in Shakelert by 2022, thus completing the system for the U.S. west coast.

3 Operational Earthquake Forecasting in Japan and the US West Coast

Earthquake predictions and forecasts are statements about future earthquake activity in a particular region. Although in common English usage, "forecast" and "prediction" are synonyms, these terms have distinct meanings in earthquake science. A prediction is an assertion that one or more earthquakes will occur in a specified location within a specified future time interval. Forecasts specify the probabilities that earthquakes will occur in the specified region and time interval. A time-independent forecast, sometimes called a "long-term" forecast, is one in which probabilities depend only on the rates at which earthquakes have occurred during a lengthy past time period; such forecasts are useful for long-term seismic hazard analysis. In contrast, in a time-dependent forecast, the probabilities are updated as new information becomes available. Earthquake forecasts may be updated in response to the discovery of previously unknown past events, in response to new research findings about fault behavior, or (slowly) as increasing time passes since the last significant event. "Operational earthquake forecasting" (OEF) refers specifically to updating of earthquake forecasts in near real time as earthquake activity occurs. In both Japan and the US, earthquake predictions and long-term forecasts have been developed by scientists and government agencies since the 1970s. OEF is a newer concept that gained prominence following the 2009 L'Aquila, Italy earthquake.

3.1 Earthquake Forecasting in Japan

The Tokai region, an area of central Japan between Nagoya and Tokyo, had experienced great earthquakes (~M8) in 1854 and 1707, and seismologists concluded that the recurrence interval for major Tokai earthquakes was approximately 120 years. As a result, an earthquake of magnitude 8 has been anticipated for the Pacific Coast of Shizuoka Prefecture, including Suruga Bay (Rikitake 1979). In an early description of the Tokai "prediction" and its legal foundation, Prof. Tsuneji Rikitake, writing in EOS (Rikitake 1979), introduced English-speaking audiences to the Large-Scale Earthquake Countermeasures Act of 1978. The Act had passed during the 84th session of the Japanese Diet in June 1978 and was enacted in December of the same year. It was first introduced to a non-Japanese scientific audience at an April 1979 UNESCO Symposium on Earthquake Prediction. Prof. Rikitake remarked in the EOS article that: "the act attracted much attention of the audience, which included many sociologists" (op cit 1979). As a sociologist, the first author of this chapter is hardly surprised at this observation given the interest social scientists have in risk communication and, more generally, human behavior during disasters. The Act was passed at a time when optimism that reliable scientific methods of predicting earthquakes were close to being realized. Chinese scientists had successfully predicted a major earthquake in Haicheng just five years earlier (Bolt 2006).

The Large-Scale Earthquake Countermeasures Act has as its objective the mitigation of earthquake hazards in the Tokai region (Rikitake 1979). According to the Countermeasures Act, the Tokai Region was to be designated "an area under intense measures against earthquake disaster" (op cit, p. 553). These measures included a dense observation network consisting of "an array of sea bottom seismographs, tiltmeters, volume strainmeters, tide gauges, and instruments for monitoring groundwater level and radon contents..." Data from this array were to be telemetered to the JMA in real time. If this network revealed anomalies believed to be consistent with the imminent occurrence of a major earthquake, as assessed by the Prediction Council (a suborganization of the Coordinating Committee for Earthquake Prediction), a warning would be conveyed to the Prime Minister via the Director General of the JMA. Upon receiving such a warning, the Prime Minister could exercise emergency powers granted by the Countermeasures Act to initiate actions that could mitigate the hazards that a major earthquake in this heavily populated region would pose.

The Act does not specify the length of a warning period other than designate it "short-term". In the EOS article, Prof. Rikitake indicates that the warning period would range from "several hours to a few days" (Rikitake 1979, p. 555). Included in the legislation is a mandate for creating the Central Disaster Prevention Council, chaired by the Prime Minister, to formulate a basic plan for response. This mandate also covered prefectural and municipal governments, and extended to the private sector including hospitals, hotels, department stores, bus and rail transportation companies, refineries and other facilities, requiring them to develop short-term disaster prevention plans within six months of enactment of the law. The Act gave the Prime Minister emergency powers to set up and direct a "national headquarters for earthquake disaster prevention". Prefectural and municipal headquarters would also be set up under a warning issued by the Prime Minister. The emergency powers granted by the Countermeasures Act allow the Prime Minister to mobilize the Self-Defense Forces and order them to enter the region prior to an anticipated earthquake. The Act also empowers prefectural and municipal authorities to take actions such as ordering shutdowns of rail transportation and oil refinery operations, regulating highway traffic, and taking over privately owned resources for emergency use. At the time of this writing, such an earthquake emergency for the Tokai region has never been declared.

Although the 1970s were a time of optimism regarding the future of earthquake prediction, this optimism was not shared by all or even most earth scientists. From the outset, the Countermeasures Act was controversial. An editorial in the Japan Times (2016) noted that the National Land Agency (one of the predecessors of the Land, Infrastructure, Transport and Tourism Ministry) that was in charge of disaster prevention at the time, was reluctant to implement the law because of the lack of scientific evidence that short-term earthquake prediction was possible. Perhaps the most outspoken critic of the Act and its scientific foundation has been Prof. Robert Geller of Tokyo University. In the "Comment" section of Nature published in April 2011, Geller argued, "it is time to tell the public frankly that earthquakes cannot be predicted, to scrap the Tokai prediction system and to repeal the LECA (Large-Scale Earthquake Countermeasures Act)." In this article, Geller attacked the basis for the Tokai prediction, stating that the characteristic earthquake model and its derivative, the seismic gap hypothesis, had failed to identify the source zones of most large earthquakes that had occurred after enactment of the law. He relates that since 1979, the earthquakes that caused 10 or more fatalities in Japan, including the Great East Japan Earthquake and Tsunami of 2011, had occurred in areas regarded as having a low probability of damaging earthquake occurrence. Geller speculated that the Tokai prediction had few critics among mainstream Japanese seismologists due to cooptation based on funding and committee memberships, lack of criticism by panels appointed to review government decisions, disregard by the media of critics of government decisions, and inadequate scientific training of reporters (Geller 2011).

Despite early and subsequent criticisms, the Japanese government has pressed ahead with the intensive monitoring program and the mandate for planning a response to a major earthquake in the Tokai area. As scientific concern grew about the hazard posed by other large subduction zone earthquakes in the Japan Trench including an extension of the Tokai event to include Nankai and Tonankai, the Japanese Diet passed an additional law designed to address the threat posed by large potentially tsunamigenic earthquakes in the Nankai and Tonankai regions. In 2002, the Japanese Diet passed the Act on Special Measures for Promotion of Tonankai and Nankai Earthquake Disaster Management. This legislation gave the Prime Minister the power to designate areas as "countermeasures promotion areas" where disaster prevention measures must be implemented due to the significant likelihood of damage in the expected Nankai and Tonankai earthquakes. Local governments in the areas identified by the Prime Minister were required to develop response plans and received financial support from the national government to do so (Umeda 2013).

Subsequent legislation and national policy required the government to enhance seismic monitoring in the Nankai and Tonankai regions. Various working groups under the national government's Central Disaster Management Council have considered the risk of a very large earthquake in the Japan Trench and the problems associated with high probability forecasting based on possible precursory seismic activity. In a report released in September 2017, the Council, acknowledging the difficulty of predicting a major earthquake, called for a review of the measures that could be taken under the 1978 Countermeasures Act, which had been passed at a time of high confidence that earthquake prediction would be possible. In the interim, while this review is underway, the JMA has launched a system in which some procedures followed in the 1978 Countermeasures Act would be implemented, including intensified seismic monitoring of the Nankai and Tonankai region. If anomalous activity is observed, a committee of experts will assess the situation and if members agree that the probability of a major earthquake has increased, a warning will be issued to people and organizations in the region, but major declarations and measures permitted under the 1978 Act will not be implemented. The current policy identifies potential precursors as a M7 or greater earthquake in the identified source region, unusual changes in strainmeter measurements following a M6 earthquake, or significant changes in strainmeter measurements without an earthquake (Umeda 2013; The Japan Times 2017).

Thus, it appears that until there is clear direction based on new legislation and/or directives from the Central Disaster Management Council, the JMA will respond to anomalous seismic activity in the Japan Trench by issuing an earthquake advisory, but not by implementing extreme measures like evacuations from vulnerable buildings, suspending railway operations, closing schools, or mobilizing the Self-Defense Forces, though these powers continue to be available to the Prime Minister should he chose to enact them under the 1978 Large-Scale Countermeasures Act, which has not been modified or suspended. The long-term forecast for a M8-9 earthquake in the subduction zone off the Pacific Coast of Japan is estimated to be 70% in the next 70 years (Japan Times 2017).

3.2 Earthquake Forecasting in the US

The optimism of the 1970s regarding the feasibility of reliable short-term earthquake prediction energized the US earthquake science community. Since the 1960s, US earthquake scientists had been working slowly toward the goal of a federally funded program for earthquake hazards reduction. Although the needs for engineering research and mitigation efforts were appreciated and advocated, Congress finally passed the National Earthquake Hazards Reduction Act in 1977 only after emphasis had been placed on the possibility of predicting earthquakes (Wallace and Scott 1999). Funding appropriated under the Act was used to deploy equipment for monitoring seismicity, crustal deformation, and other geophysical and geochemical parameters that had been alleged to have predictive value for earthquakes. The instrumentation was focused on the San Andreas fault system in California.

Of particular note was the initiation of the Parkfield Earthquake Prediction Experiment. Parkfield, on the San Andreas fault about halfway between San Francisco and Los Angeles had experienced earthquakes of about M6 in 1857, 1881, 1901, 1922, 1934, and 1966. On the basis of this sequence of events, Bakun and Lindh (1985) estimated the recurrence interval for magnitude 6 earthquakes near Parkfield as 22 $\pm$ 3 years and conjectured with 95% confidence that another such event would occur before 1993. After Bakun and Lindh's prediction was accepted by the National Earthquake Prediction Evaluation Council (NEPEC) and the California Earthquake Prediction Evaluation Council (CEPEC), dense observation networks were established along the Parkfield fault segment. The experimental goals were to record geophysical signals accompanying the earthquake process at Parkfield, to capture effects of strong motion in the near field, and, most ambitiously, to attempt to issue a warning up to 3 days before the next M6 earthquake if seismicity and/or fault creep rate reached predefined threshold levels (Roeloffs and Langbein 1994). To prepare the local population for the possible issuance of a pre-earthquake warning, the California Office of Emergency Services devised a response plan and disseminated pamphlets describing actions that could be taken in the affected area. Although the anticipated Parkfield earthquake did not occur until 2004 (Bakun et al. 2005), alerts were issued twice, and social science studies showed that local residents took preparedness actions in response to those alerts (Mileti et al. 1992).

Concurrent with the Parkfield experiment, earthquake scientists in the US were being urged to quantify estimates of the probability of large, damaging earthquakes in California based on the available knowledge regarding past earthquake behavior. As such, scientists were charged with making long-term forecasts, taking into account past behavior and updating in response to new information, new understanding of the earthquake process, the occurrence of major earthquakes, and, eventually, the passage of time. To address this need, the 1988 Working Group on California Earthquake Probabilities was formed and produced a report (WGCEP 1988) that estimated probabilities that large earthquakes ($M \geq 7$) would occur on parts of the San Andreas fault system during the 30-year period ending in 2018. Inherent in the methodology of this report was the assumption that earthquakes recur on fixed fault segments whose boundaries could be inferred from the extent of the rupture during previous events. Limitations of this assumption became apparent when the 1989 M_w6.9 Loma Prieta earthquake occurred on a fault adjacent to, but not coincident with, the southern Santa Cruz Mountains segment of the San Andreas fault. To update the probability estimates following the Loma Prieta earthquake, a second working group was formed in 1990, which updated the WGCEP (1988) probabilities using refined assumptions about segment boundaries, some updated slip rates, and a somewhat broader suite of assumptions implemented using logic tree analysis. A third working group report (WGCEP 1995) was developed after the 1992 Landers earthquake had demonstrated the capability of an earthquake to rupture multiple fault segments.

All of the WGCEP forecasts required not only updates but also methodological revisions after significant earthquakes (Field 2007), and as a consequence, successive WGCEP probability estimates did not always agree. For example, a salient, though controversial, result in WGCEP (1995) was that M6-7 earthquakes would occur during the forecast period at a rate twice as high as in the historical record (Field 2007). Earthquake probabilities based on WGCEP forecasts, and especially changes in those forecasts with successive updates, generated media attention (e.g., Prado 2016).

Also in the 1980s, the state of California adopted an approach to short-term earthquake forecasting that was completely separate from the Working Group process. Under this approach, which was begun in 1985, the California Office of Emergency Services (Cal OES) has issued advisory statements to local jurisdictions and the public following seismic activity that scientists on the California Earthquake Prediction Evaluation Council view as indicating elevated probability of a larger earthquake during the next several days. These advisory statements are motivated by statistical studies showing that about 5% of moderate earthquakes in California are followed by larger events within a 10 km, 5-day space–time window (Jones 1985; Agnew and

Jones 1991; Reasenberg and Jones 1994). Cal OES issued four earthquake advisories from 1985 to 1989.

In October 1990, the California Earthquake Advisory Plan formalized this practice, and six Cal OES Advisories have been issued since then. Roeloffs and Goltz (2017) present a history of the Advisory Plan, which provides for messaging to the public recommending basic preparedness actions, including knowing how to respond to earthquakes in various locational contexts. Local emergency services organizations and first responders are advised to consider such actions as reviewing earthquake response plans, partially activating emergency operations centers, reviewing staffing for the Advisory period to ensure adequate levels and types of personnel, and assuring that certain equipment (e.g., emergency power generators) is functioning properly. At this writing, no larger earthquakes have occurred during the periods of the advisories that have been issued, and little research has been carried out to evaluate the Plan's effectiveness in promoting earthquake safety. Nevertheless, we speculate that the existence of the Plan might have spared California from a situation like that of the 2009 L'Aquila, Italy, earthquake, in which public officials mistranslated seismologists' cautionary advice regarding the possibility of a damaging earthquake occurring following a vigorous seismic swarm (Hall 2011; Jordan 2013).

Following the L'Aquila debacle, an international commission of earthquake scientists was convened, and recommended that earthquake scientists develop objective methods to continuously estimate probabilities of damaging earthquakes, providing the results of these analyses to stakeholders who could decide for themselves how to use the information (Jordan et al. 2011). The process of computing continuously updated earthquake probabilities as seismicity (or, in some cases, crustal deformation) is observed was named "Operational Earthquake Forecasting" (OEF). OEF differs from periodically updated forecasts like those of the WGCEPs in that all seismicity affects probabilities, rather than only certain events that earthquake scientists identify as significant. At present, OEF programs are operating in Italy (Marzocchi et al. 2014) and New Zealand (Rhoades et al. 2016). Here, we describe the approaches to OEF that are being developed and evaluated in the US.

OEF updates earthquake forecasts in response to seismic activity by assuming that future earthquakes will follow the patterns of clustering and triggering that have been deduced from statistical studies of earthquake catalogs, and/or that are plausible based on current scientific understanding of how active faults behave. As such, we believe that OEF is a natural extension of aftershock forecasting. The most robust statistical relationships are found between an earthquake and subsequent events (aftershocks) that occur within a short distance (order of kilometers) and a short time (seconds to days): within this narrow time–space window, earthquake probabilities can temporarily increase by several orders of magnitude before decaying to the background rate. The earliest studies of such clustering and triggering relations (e.g., Jones 1985; Reasenberg and Jones 1989; Agnew and Jones 1991) considered only the probability that an individual earthquake would be followed by one or more additional events in a given magnitude range. Subsequent statistical studies have validated the principle that every earthquake can trigger additional nearby earthquake sequences, known as "epidemic-type aftershock sequences", referred to as the "ETAS" model, and originally developed in Japan by Ogata (1988). The original ETAS model provides only estimates of the numbers of triggered events of a given magnitude as time elapses after the triggering earthquake, but has been extended to forecast the distribution of events with respect to distance from the triggering event (Ogata 1998; Ogata and Zhuang 2006).

All statistical models of earthquake triggering and clustering include adjustable parameters that must be fit to the available data on earthquake occurrence for a particular region, and that often also require adjustment as an earthquake sequence evolves. These parameters include the overall rates of triggered events, and the rapidity with which they decay with distance and elapsed time from the triggering earthquake. Currently, the Reasenberg and Jones (1989) formula is routinely used to forecast the aftershock sequences of M4 and larger earthquakes in California, where earthquake catalogs are adequate to define good starting estimates for model parameters. The Reasenberg and Jones (1989) model is currently being implemented for nationwide earthquake forecasting (Page et al. 2016), albeit with a higher magnitude threshold outside of California, owing to less complete information on past seismicity. However, because the Reasenberg and Jones (1989) formulation does not take into account events triggered by aftershocks themselves ("secondary aftershocks"), it can require continual parameter adjustment as a sequence evolves. This model is also of limited usefulness in forecasting sequences in regions of suspected induced seismic activity (Michael et al. 2016; Llenos and Michael 2013). Current plans in the US call for aftershock forecasting to transition to an ETAS-based model as soon as practicable. To "operationalize" earthquake forecasting, a simple approach is to continuously calculate earthquake probabilities using the ETAS formulation; this approach is likely the first that will be deployed in the US.

To improve on the success rates of forecasts, researchers have identified other subtle behavior patterns in earthquake catalogs, and several such forecasting methods are under long-term prospective tests in the Collaboratory for the Study of Earthquake Predictability (CSEP) (Zechar et al. 2010; Strader et al. 2017). A subset of these forecasts modify probabilities in response to the stress changes that earthquakes presumably impose on nearby faults; incorporating this spatial dimension requires knowledge of fault geometry, earthquake depths, and focal mechanisms. An important forecast model for California is the Uniform California Earthquake Rupture Forecast, Version 3 (UCERF3; Field et al. 2014, 2015), which generates earthquake probabilities from sets of possible future events computed such that constraints including seismic moment balance, historic and prehistoric event rates, and elastic rebound are satisfied, while allowing for the possibility of multi-segment ruptures (see Fig. 5). Field et al. (2017) describe an enhancement of UCERF3 to include an ETAS clustering model that would arguably represent the most sophisticated approach to OEF that could be implemented in California at present. Prospective testing will be required to evaluate whether UCERF3-ETAS can outperform less computationally intense forecasting methods.

All present OEF methods generate nonzero probabilities of a much larger future event, but such probabilities are very small and decrease within hours to days of

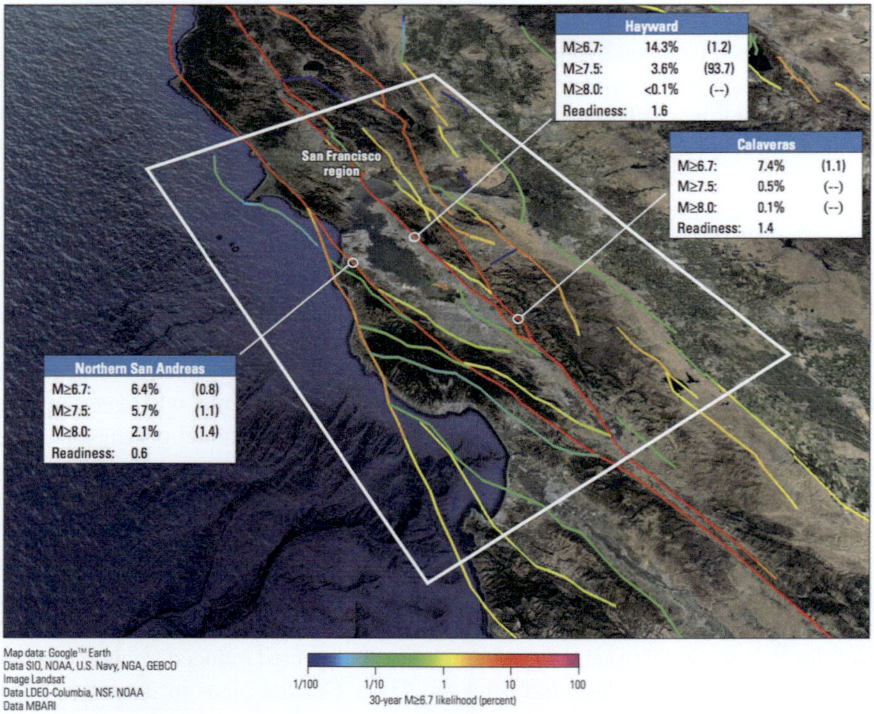

Fig. 5 UCERF3 likelihood of magnitude 6.7 or greater earthquakes in the next 30 years, from 2014, on the faults near San Francisco, Calif. Courtesy of the United States Geological Survey, US Department of the Interior (pubs.usgs.gov/fs/2015/3009/pdf/fs2015-3009.pdf). Accessed 9 Jan 2018

triggering events. The narrowness of this time window means that these OEF methods can provide only very short term or imminent warnings. The low background probabilities of most earthquakes mean that the absolute probability of a damaging event in this time–space window is still very low: usually about 5% in a week, and only rarely above 25%. Advance preparation is required in order to use this type of forecast to increase safety and reduce losses in earthquakes. In the U.S., the California Earthquake Advisory Plan is the only plan that delineates actions to be taken by the state emergency management agency when a heightened short-term earthquake probability is forecast.

Yet California is not the only region of the US with high earthquake hazard. For example, in the last 25 years, consensus has formed that the Cascadia Subduction Zone (CSZ), which extends from northernmost California to Vancouver Island, Canada, has generated earthquakes as large as M9 at intervals of several hundred years (see, for example, Frankel et al. 2015). The next great CSZ earthquake thus threatens the states of Oregon and Washington, in addition to California and part of Canada's western coast, with not only powerful shaking, but also extreme coastal tsunami hazard. Developing quantitative short-term forecasts for Cascadia earthquakes is problematic (Gomberg et al. 2015). Background seismicity is much quieter in Cascadia than in California, so that seismic monitoring has been less intensive and has operated for a shorter time. As a result, earthquake catalogs for the Cascadia region do not permit the robust derivation of statistical parameters that is possible in California, especially not for a major CSZ earthquake, which last occurred in 1700 CE. At the same time, the deployment of continuous GPS stations and borehole strainmeters throughout Cascadia has revealed that aseismic slip events recur repeatedly at intervals of months to a few years in the deeper portions of the CSZ. Such events have been hypothesized to be capable of triggering major subduction earthquakes (e.g., Segall and Bradley 2012), but at present, there is no realistic method for incorporating aseismic slip observations into probability calculations. Lacking a protocol like the California Earthquake Advisory Plan, there is currently no procedure in the Cascadia region by which earthquake scientists could evaluate the potential for unusual seismicity or aseismic slip to indicate heightened short-term probability of a major Cascadia earthquake. Motivated by this need, a process was initiated in 2016 to solicit information from stakeholders, including state and local emergency managers, regarding their preferences for how earthquake scientists should convey their conclusions. The consensus report of that process supports the formation of a Cascadia-wide council analogous to CEPEC, which would evaluate observations and arrive at recommendations as to whether an Advisory-type message is warranted.

4 Discussion: Anticipating Large Earthquakes and Their Effects in Japan and the US

At the outset of this chapter, we stated that earthquake prediction, a deterministic statement that a future earthquake will or will not occur in the narrowly defined context of time, location, and magnitude range (Jordan et al. 2011), is not currently possible. Thus, EEW and low-probability short-term forecasts are the scientific reality, for now. But these technologies are important and could contribute to earthquake hazard mitigation and reduced casualties in a major earthquake in the United States and Japan. Although in its conceptual form prior to implementation, EEW had critics, there is now consensus both in Japan and the United States that this technology has value and should receive resources for further development and implementation. OEF, on the other hand, remains controversial, both as an emerging technology and as a useful public good. In the remainder of this section, we will discuss the issues involving low-probability earthquake forecasts, outline potential benefits of implementing OEF in the US and Japan, and point out how OEF could enhance EEW's effectiveness.

Proponents of OEF argue that it is appropriate for residents of seismically vulnerable regions and their institutions to expect that scientists will not withhold from them authoritative information regarding time-dependent probabilities of a damaging earthquake that may affect them. The risk information communicated based on short-term forecasts, and accompanying recommended preparedness actions, must be consistent with the low probability of occurrence. That is, large-scale movements of resources, significant disruptions of economic and commercial activity, or evacuations of vulnerable buildings may be unwarranted, though communication to populations in the region of ongoing seismic activity that risk has increased over a few days of duration is certainly justified in the interests of transparency.

Critics of OEF claim that warnings based on very low probabilities send the wrong message to the public, that if earthquake probabilities on a daily basis are negligible (0.0001%) and a "heightened" daily probability is just 5%, then the public will perceive that it can be less prepared for earthquakes. Further, they argue, an increase in short-term probability from "practically impossible (0.0001%) to very unlikely (5%) is not useful for saving lives" (Wang and Rogers 2014). Interestingly, these arguments are based, not so much on scientific evidence but assumptions about societal response to warnings. These critics suggest that, "crying earthquake is a potent way of blunting earthquake awareness and preparedness" and that short of providing guidance in the evacuation of vulnerable buildings, a short-term earthquake forecast that advocated partial measures where life and death may be the consequences "would trigger runaway reactions among the public, especially in the age of social media" (op cit., p. 570). These authors reason that instead of short-term forecasts, the public policy conversation should focus on long-term structural hazard mitigation, particularly the mitigation of hazards associated with vulnerable structures.

While we agree that mitigation of seismically vulnerable buildings should be a long-term hazard reduction priority, we argue that short-term forecasts also have value for public safety. Proponents of short-term earthquake forecasting (Jordan et al. 2011, 2014; Jordan 2013; Goltz 2015; Field et al. 2016; Roeloffs and Goltz 2017) have pointed out a number of potential uses for OEF and justifications for its implementation as a risk mitigation measure. Despite operating in a low-probability environment for large damaging earthquakes, short-term OEF can provide high probability forecasts for aftershocks that are felt but cause little damage, information which can be reassuring in the aftermath of a larger event (Wein and Becker, 2013). With a low-probability forecast, there are useful alternatives to large-scale evacuations of people from vulnerable buildings (and many of the measures envisioned under the 1978 Earthquake Countermeasures Act in Japan). Simple low impact recommendations to the public that they reinforce preparedness by storing water, extra food and medical supplies, agree on a family emergency plan, and secure furniture and movable objects can be made during the warning period and will continue to promote earthquake safety after the warning expires (Goltz 2015; Field et al. 2016).

For some organizations such as utilities, critical facilities, and transportation lifelines, short-term low-probability forecasts may have significant value when risk tolerance is low. For these, as well as other organizations and the public in general, short-term forecasts can reinforce other loss prevention measures including earthquake early warning systems, when available in a region. A utility, for example, may be capable of implementing mitigation measures in response to a short-term, low-probability forecast that would not be feasible in the imminent time frame of an EEW alert. Such capabilities could, at minimum, be reinforced through review and simulation or exercise during a short-term forecast. Critical facilities like hospitals can postpone high risk and elective surgeries, ensure that vital equipment is secured and that backup power is available and functional (Field et al. 2016). For all potential users of short-term forecasts, the time period for which a population is warned of a possible large earthquake constitutes a "teachable moment" in which preparedness is enhanced, rather than blunted as critics have charged (Op cit., p. 316).

Short-term OEF also has had utility for emergency management organizations and units of government that serve as first responders in an emergency. Since 1990, the state of California has had an "earthquake advisory plan" (California Office of Emergency Services 1990), which is essentially a short-term OEF plan. As described earlier and in Roeloffs and Goltz (2017), state officials have exercised this plan multiple times, preparing emergency management organizations for possible damaging earthquakes without provoking counterproductive public actions.

An underlying theme and critical component in any warning communication is the message itself. Social scientists have long emphasized that the communication of risk is ineffective unless concrete measures are recommended to mitigate personal and collective risk (Mileti and Sorenson 1990; Mileti 1999; Field et al. 2016). Social science research has also debunked the conventional wisdom that the public is likely to respond to warning messages in a maladaptive manner as suggested by some critics of OEF, especially if they are properly informed of risks and advised on preparedness and mitigation actions (Mileti 1999; Goltz 2015). In the situation of L'Aquila, often cited by proponents of OEF, poor communication between scientists and emergency managers, inconsistent and contradictory information provided to the community and belated deterministic, rather than probabilistic statements about the imminent occurrence of an earthquake characterized a tragic chain of consequences that spoke eloquently to the need for planning and scientific acceptance of the obligation to speak with authority on natural hazards, including the communication of the uncertainty regarding the probability of occurrence (Goltz, op cit., p. 1232).

In summary, we characterize short-term earthquake forecasts and earthquake early warning as limited but useful technologies that are relevant to the pre-earthquake period as means of preparedness and hazard mitigation. But these limitations do not negate the advantages these technologies offer in terms of life safety and hazard mitigation as elaborated in this section. Though limited, these technologies are already available to be incorporated into the loss prevention arsenals and long-term strategies of hazard reduction. In the final section, we will offer some ideas on how these technologies might be implemented in Japan and the United States.

5 Conclusions and Recommendations

We would like to return to the statements of Robert Geller from his post-Tohoku Comment in the April 2011 edition of *Nature*, where he states:

> All of Japan is at risk from earthquakes, and the present state of seismological science does not allow us to reliably differentiate the risk level in particular geographic areas. We should instead tell the public and the government to 'prepare for the unexpected' and do our best to communicate both what we know and what we do not.

We agree with this statement and place particular emphasis on doing our best to communicate what we know and what we do not. Telling the public in simple straightforward language what scientists know and do not know is, or should be, the base from which risk is communicated. And we interpret Geller's statement to imply that during an evolving seismic sequence, or the observation of highly anomalous monitored data, scientists should tell the public everything known about the events including the seismic history of the region experiencing the activity. To say nothing would be irresponsible; it might also create an information vacuum into which unqualified people can make unsubstantiated statements about what the situation may portend.

Japan's legislated earthquake countermeasures act currently permits the Prime Minister to take many actions whose benefits are likely outweighed by their potential for economic and other types of disruption. Nevertheless, the act also mandates various levels of government and the private sector to develop plans for response to major earthquakes in the Tokai, Nankai, and Tonankai regions, and requires intensified seismic monitoring in certain areas with high seismic risks to large populations and critical infrastructure. We acknowledge that short-term high probability statements about earthquake occurrence are unwarranted, but we see it as important to understand the limitations and challenges while pursuing the benefits of OEF for all of Japan and for the region of the West Coast of the United States adjacent to the Cascadia Subduction Zone.

Important factors are the seismic vulnerability of Japan and that nation's lengthy history of large highly damaging earthquakes with staggering losses of life. Geller (2011) and others in their critiques of the Tokai "prediction" have pointed out that since the Large-Scale Earthquake Countermeasures Act was implemented in 1979, earthquakes that have resulted in fatalities have occurred in areas of Japan assigned relatively low probabilities. But this correct observation can alternatively be seen, not as a reason for retreat from forecasting, but rather as a reason to expand forecasting to all of Japan. In a nation in which an earthquake-induced tsunami can claim the lives of 18,000 people, create a nuclear disaster, and cause losses in excess of $220 billion, the benefits of implementing OEF to improve preparedness may outweigh the costs. Even slight increases in the localized probability of damaging earthquakes could be opportunities to advise the public to be cognizant of personal safety in an earthquake, to reinforce the need to prepare, and to promote measures that would limit damage and maintain safety.

The two most recent and destructive earthquakes in Japan were preceded by foreshocks that would have caused typical OEF algorithms to raise estimates of the probabilities of larger earthquakes in the next several days. Prior to the Mj9.1 Great East Japan Earthquake and Tsunami, there was a pre-shock of M7.4 on March 9 followed by three M6 events on the same day (Hayes et al. 2017). Nanjo et al. (2012) show retrospectively that these events elevated the short-term probability of a larger event around the epicenter of the Mj9.1 earthquake, as calculated for March 11,

2011, by two forecasting algorithms under test in the CSEP. In addition, retrospective analyses of displacements measured by the 1240-station, Japan-wide GEONET (GPS Earth Observation Network) array identified anomalous crustal movements consistent with slip-rate changes on the Japan Trench plate interface (Mavrommatis et al. 2014), 3 days (Kamiyama et al. 2012), and 50 h (Miyazaki et al. 2011) prior to the Mj9.1 earthquake. Kamiyama et al. (2012) pointed out that had the shorter-term crustal movements occurred in the Tokai region, where they would more likely have been detected in real time, they would have prompted a public warning and triggered the actions identified in the 1978 Countermeasures Act. Although those actions may not have been justified, some measures could have been implemented, such as a public acknowledgment that the M7.4 and crustal anomalies could indicate a temporarily elevated probability of a larger event in the next several days, with reminders to residents and visitors in the coastal areas of Tohoku to be aware of the EEW system, to be prepared for further earthquakes, and to remember that a large subduction earthquake could generate a fast-arriving tsunami, making it prudent to review evacuation routes.

On April 14, 2016, a Mj 6.5 earthquake occurred at 9:26 PM JST in Kumamoto Prefecture Japan. Between 10:07 PM and 11:43 PM on April 14th, there were three $M \geq 5$ events and at 12:03 AM on the 15th an aftershock of M6.4 occurred. At 1:25 AM JST on April 16, 2016, a larger earthquake occurred which was recorded by the JMA at Mj7.3 (USGS M7.0) located at latitude 32.8 and longitude 130.7E at a depth of 10 km. It was located 1.0 km (0.6 mi) WSW of Kumamoto City. Like its foreshock, this earthquake had a strike-slip focal mechanism and occurred at shallow depth. JMA reported a maximum JMA Intensity of 6-Upper. A total of nine people died in the April 14th event and 41 in the Mj7.3, 28 h later. It is likely that more people would have died in the mainshock had there not been many people in emergency shelters due to damage to homes and multifamily buildings in the initial earthquake on April 14th (Nakashima et al. 2017). It is possible that more lives may have been spared had there been an operational earthquake forecasting system in place. At the very least, those who had evacuated after the first event might have been cautioned to remain in safe structures for 3–5 days rather than return to damaged homes, to be aware of Japan's earthquake early warning system, and to be prepared to take action when alerted to further earthquake activity.

In the U.S., the Cascadia Subduction Zone, extending from northern California to Vancouver Island, Canada, has a 10,000 year record of producing earthquakes of M8.5 or greater at intervals of several hundred years, with the last major event (likely a magnitude 9) in 1700. The next major Cascadia event will produce several minutes of strong shaking in northern California, western Oregon and Washington, and British Columbia, accompanied by a tsunami that will impact many coastal communities and that will arrive in tens of minutes. The planning scenario must acknowledge that a catastrophic Cascadia earthquake could occur without warning. However, geodetic and seismic monitoring capabilities have improved, and could now detect phenomena in Cascadia that might indicate that the probability of a dangerous event was temporarily elevated. Examples include a significant change in the seismicity of the subduction interface, or the occurrence of aseismic slip events that exceed the size of those now known to regularly occur. The majority of the subduction interface has remained seismically quiet at about the magnitude 5 level for the duration of modern seismic recording, so that any seismicity increase on the subduction interface, say at the M6 level, would come under scrutiny as a possible foreshock. Indeed, if such an earthquake were to occur now, California's Office of Emergency Services would issue a public earthquake advisory.

There is also a need to anticipate a wider variety of possible scenarios than just "The Big One", for which area residents are typically urged to prepare. For example, paleoseismic data indicate that past great Cascadia earthquakes have had a range of magnitudes. This finding means that the entire zone could rupture sequentially in a series of M8-8.5 earthquakes; immediately after the first such shock, seismologists and geodesists would be expected to intensively investigate the expected time frame of the next earthquake in the sequence, a task that would be far more tractable with an OEF system in place for Cascadia. The relationship of aseismic slip events in Cascadia with the time-dependent earthquake hazard is an active topic of research, now facilitated by a large supply of data from continuous GPS stations and strainmeters of the Plate Boundary Observatory. If methods of including aseismic slip information in OEF forecasting methods are developed, emergency managers can determine how to take advantage of such information.

We also see the potential for earthquake early warning and short-term operational earthquake forecasting to be implemented in a synergistic manner. Both are products of seismic monitoring and both can be considered real-time information products. On the most basic level, in an area or region having a scientifically assessed short-term elevated seismic risk, if public information can highlight the possibility that an early warning may be issued, residents of the area can be encouraged to review actions that can be taken to assure personal safety in a variety of locational contexts should an EEW message be received. In addition, based on the short-term forecast, loss estimates can be calculated for an event that may occur and residents of the area under an advisory would be better informed of the level of ground motion they may expect if an early warning message is issued. The two technologies, operating together may have psychological advantages. For residents and visitors in an area under a short-term forecast, knowledge that they are likely to receive advanced warning if an earthquake occurs would provide some level of comfort and reassurance.

Cauzzi et al. (2016) have suggested how OEF and EEW could contribute synergistically to mitigation of risk at nuclear power facilities. These authors write that model forecasts based on a STEP algorithm (Gerstenberger et al. 2005) can be utilized to compute time-varying daily earthquake rate forecasts and probabilities of ground shaking throughout the region in which the nuclear power facility is located. These model forecasts expressed in a mapped format can be updated on an hourly basis in situations in which there is an ongoing earthquake sequence. Nuclear facility control room staff can be trained to use the information in such maps to be aware of the seismic threat level and enhance preparedness for possible near future ground shaking. OEF could even be incorporated into EEW algorithms, providing a scientifically plausible first guess at earthquake location and magnitude, potentially speeding algorithms' convergence to correct source characterization (Cauzzi et al.

2016). These authors also mention other preparedness and mitigation measures that could be implemented at nuclear facilities based on EEW and OEF including early activation of emergency plans, automatic opening of on-site fire station doors, reinforcing inspections, conducting earthquake drills, anticipating the outage period, and reducing the number of nonessential personnel on site.

These measures highlight the need to gear preparedness and mitigation actions to both the strengths and weaknesses of EEW and OEF. In adapting these technologies, it is important that the principles of transparency and hazard-risk separation be followed. The former states that: "authoritative scientific information about future earthquake activity should not be withheld from the public" (Jordan and Jones 2010) and the latter affirms that this information "should be developed independently of its applications to risk assessment and mitigation" (Jordan et al. 2014). The hazard-risk separation principle means that scientists and emergency managers must stay in their own lanes: scientists are responsible for the risk assessment and emergency managers for risk analysis and mitigation.

Japan and the United States have done much to prepare and mitigate the impacts of large earthquakes and tsunamis and both nations have embraced EEW, but more is possible. Japan's much-beleaguered Large-Scale Earthquake Countermeasures Act includes constructive actions, such as mandating planning and seismic monitoring. The recent plans (in California, this plan is called the California Cascadia Subduction Zone Earthquake and Tsunami Response Plan; Washington and Oregon have similar plans) developed to respond to a major Cascadia earthquake and tsunami could incorporate plans to issue advisories should identifiable precursors be observed in the Cascadia Subduction Zone. Both nations can benefit from OEF, developing plans to adequately inform the public, and recommending actions that are consistent with low-probability forecasts. The measures that we have reviewed in this chapter seem reasonable and appropriate for the state of forecasting technology and could save lives in the next large subduction zone earthquakes. The potential for use of EEW and OEF technologies in tandem is also significant and represents an opportunity to reduce earthquake losses in Japan and the US as well as in other nations that are at a state of development to implement them.

References

Agnew DC, Jones LM (1991) Prediction probabilities from foreshocks. J Geophys Res 96:11959–11971

Allen RM, Gasparini P, Kamigaichi O, Böse M (2009) The status of earthquake early warning around the world: an introductory overview. Seismol Res Lett 80(5)

Allen RM, Given DD, Heaton TH, Vidale JE (2014) Successful ShakeAlert performance for the Napa quake, Abstract S44D-01 presented at 2014 Fall Meeting, AGU, San Francisco, Calif., 15–19 December

Bakun WH, Aagaard B, Dost B, Ellsworth WL, Hardebeck JL, Harris RA, Ji C, Johnston MJS, Langbein J, Lienkaemper JJ, Michael AJ et al (2005) Implications for prediction and hazard assessment from the 2004 Parkfield earthquake. Nature 437(7061):969–974

Bakun WH, Lindh AG (1985) The Parkfield, California, earthquake prediction experiment. Science 229:619–624

Bakun WH, Fischer FG, Jensen EG, VanSchaack J (1994) Early warning system for aftershocks. Bull Seismol Soc Am 84(2):359–365

Bay Area Rapid Transit District (2017, October 18) BART to participate in the Great California ShakeOut 2017. News Release

Bolt Bruce (2006) Earthquakes: 2006 centennial update, 5th edn. Freeman, W.H

Böse M, Allen R, Brown H, Gua G, Fischer M, Hauksson E, Heaten T, Hellweg M, Liukis M, Neuhauser D, Maechling P (2014) CISN ShakeAlert: an earthquake early warning demonstration system for California. Early warning for geological disasters. Springer, Berlin Heidelberg, pp 49–69

Cauzzi C, Behr Y, Le Guenan T, Douglas J, Auclair S, Woessner J, Clinton J, Wiemer S (2016) Earthquake early warning and operational earthquake forecasting as real-time hazard information to mitigate seismic risk at nuclear facilities. Bull Earthq Eng 14(9):2495–2512

Doughton S (2017, April 10) Earthquake early-warning system comes to Washington—but it's not for the public yet, Seattle Times. https://www.seattletimes.com/seattle-news/science/earthquake-early-warning-system-comes-to-washington-but-its-not-for-the-public-yet/

Field EH (2007) A summary of previous working groups on California earthquake probabilities. Bull Seismol Soc Am 97(4):1033–1053

Field EH, Arrowsmith RJ, Biasi GP, Bird P, Dawson TE, Felzer KR, Jackson DD, Johnson KM, Jordan TH, Madden C, Michael AJ (2014) Uniform California earthquake rupture forecast, version 3 (UCERF3)—The time-independent model. Bull Seismol Soc Am 104(3):1122–1180

Field EH, Biasi GP, Bird P, Dawson TE, Felzer KR, Jackson DD, Johnson KM, Jordan TH, Madden C, Michael AJ, Milner KR (2015) Long-term time-dependent probabilities for the third Uniform California Earthquake Rupture Forecast (UCERF3). Bull Seismol Soc Am 105(2A):511–543

Field EH, Milner KR, Hardebeck JL, Page MT, van der Elst N, Jordan TH, Michael AJ, Shaw BE, Werner MJ (2017) A spatiotemporal clustering model for the Third Uniform California Earthquake Rupture Forecast (UCERF3-ETAS): toward an operational earthquake forecast. Bull Seismol Soc Am 107(3):1049–1081

Field EH, Jordan TH, Jones LM, Michael AJ, Blanpied ML, Other Workshop Participants (2016) The potential uses of operational earthquake forecasting. Seismol Res Lett 87(2A) https://doi.org/10.1785/02201501874

Frankel A, Chen R, Petersen M, Moschetti M, Sherrod B (2015) 2014 Update of the Pacific Northwest portion of the US National Seismic Hazard Maps. Earthq Spectra 31(S1):S131–S148

Fujinawa Y, Noda Y (2013) Japan's earthquake early warning system on 11 March 2011: performance, shortcomings, and changes. Earthq Spectra 29(S1):S341–S368

Geller RJ (2011) Shake-up time for Japanese seismology. Nature 472:407–409

Gerstenberger MC, Wiemer S, Jones LM, Reasenberg PL (2005) Real-time forecasts of tomorrow's earthquakes in California. Nature 435(7040):328–331

Given DD, Cochran ES, Heaton T, Hauksson E, Allen R, Hellweg P, Vidale J, Bodin P (2014) Technical implementation plan for the ShakeAlert production system—an earthquake early warning system for the West Coast of the United States: U.S. Geological Survey Open-File Report 2014–1097, p 25. http://dx.doi.org/10.3133/ofr20141097

Goltz JD (2003) Applications for new real-time seismic information: the TriNet project in Southern California. Seismol Res Lett 4(5):516–521

Goltz JD (2015) A further note on operational earthquake forecasting: an emergency management perspective. Seismol Res Lett 86(5). https://doi.org/10.1785/0220150080

Gomberg J, Atwater B, Beeler N, Bodin P, Davis E, Frankel A, Hayes G, McConnell V, Melbourne T, Oppenheimer D, Parrish J and others (2015) Earthquake forewarning in the Cascadia region. US Geological Survey Open-File Report, 1151(8)

Grapenthin R, Johanson I, Allen RM (2014) The 2014 M_w6.0 Napa earthquake, California: observations from real-time GPS-enhanced earthquake early warning. Geophys Res Lett 41:8269–8276. https://doi.org/10.1002/2014GL061923

Hall SS (2011) At Fault? Nature 477:264–269
Hartog JR, Kress VC, Malone SD, Bodin P, Vidale JE, Crowell BW (2016) Earthquake early warning: ShakeAlert in the Pacific Northwest. Bull Seismol Soc Am 106(4):1875–1886
Hayes G, Meyers EK, Dewey JW, Briggs RW (2017) Tectonic summaries of magnitude 7 and greater earthquakes from 2000 to 2015. USGS Open-File Report, 2016–1192
Heaton TH, Series N, May N (1985) A model for a seismic computerized alert network. Science 228:987–990
Hickey H (2017, April 10) USGS, partners launch a unified, West Coast-wide earthquake early warning system, University of Washington News
Holden R, Reichle M, Lee R (1989) Technical and economic feasibility of an earthquake warning system in California (Special Publication 101, California Division of Mines and Geology)
Japan Meteorological Agency (2012) Results of the survey on the utilization of emergency earthquake information. http://www.jma.go.jp/jma/press/1203/22c/manzokudo201203.html (in Japanese)
Japan Meteorological Agency (2013) Lessons learned from the tsunami disaster caused by the 2011 Great East Japan Earthquake and improvements in JMA's Tsunami warning system. http://www.data.jma.go.jp/svd/eqev/data/en/tsunami/LessonsLearned_Improvements_brochure.pdf
Japan Meteorological Agency (First Observed 2017) Emergency warning system: a new service to protect life, http://www.jma.go.jp/jma/en/Emergency_Warning/Leaflet (Emergency_Warning_System).pdf
Japan Meteorological Agency (2016) On the 2016 Kumamoto earthquake (the 39th report) (press release). http://www.jma.go.jp/ma/press/1606/10a/kaisetsu201606101000.pdf
Japan Ministry of Internal Affairs and Communications (2014) Booklet of best practices of resilient ICT systems in Japan. http://www.soumu.go.jp/main_content/000372211.pdf
Japan Times (2016, July 7) Don't rely on quake predictions, Editorial
Japan Times (2017, November 5) Anticipating a major Nankai Trough quake, Editorial
Jones LM (1985) Foreshocks and time-dependent earthquake hazard assessment in southern California. Buletin Seismol Soc Am 75:1669–1679
Jordan TH, Jones LM (2010) Operational earthquake forecasting: some thoughts on why and how. Seismol Res Lett 81:571–574
Jordan TH, Marzocchi W, Michael AJ, Gerstenberger MC (2014) Operational earthquake forecasting can enhance earthquake preparedness. Seismol Res Lett 85(5):955–959
Jordan TH, Chen YT, Gasparini P, Maderiaga R, Main I, Marzocchi W, Papadopoulos G, Sobolev G, Yamaoaka K, Zschau J (2011) Operational earthquake forecasting: state of knowledge and guidelines for implementation: final report of the international commission on earthquake forecasting for civil protection. Ann Geophys 54(4):315–391. https://doi.org/10.4401/ag-5350
Jordan TH (2013) Lessons of L'Aquila for operational earthquake forecasting. Seismol Res Lett 84(1). https://doi.org/10.1785/0220120167
Kamiyama M, Sugito M, Kuse M (2012) Precursor of crustal movements before the 2011 Great East Japan earthquake. In: Proceedings of the international symposium on engineering lessons learned from the 2011 Great East Japan earthquake, March 1–4 2012, Tokyo, Japan
Kodera Y, Saitou J, Hayashimoto N, Adachi S, Morimoto M, Nishimae Y (2016) Earthquake early warning for the 2016 Kumamoto earthquake: performance evaluation of the current system and the next generation methods of the Japan Meteorological Agency. Earth, Planets and Space 68:202. https://doi.org/10.1186/s40623-016-0567-1
Kohler MD, Cochran ES, Given D, Guiwits S, Neuhauser D, Henson I, Hartog R, Bodin P, Kress V, Thompson S, Felizardo C, Brody J, Bhadha R, Schwarz S (2017) Earthquake early warning ShakeAlert system: West Coast wide production prototype. Seismol Res Lett 89(1):99–107. https://doi.org/10.1785/0220170140
Llenos AL, Michael AJ (2013) Modeling earthquake rate changes in Oklahoma and Arkansas: possible signatures of induced seismicity. Bull Seismol Soc Am 103(5):2850–2861
Marzocchi W, Lombardi AM, Casarotti E (2014) The establishment of an operational earthquake forecasting system in Italy. Seismol Res Lett 85(5):961–969

Mavrommatis AP, Segall P, Johnson KM (2014) A decadal-scale deformation transient prior to the 2011 *Mw* 9.0 Tohoku-oki earthquake. Geophys Res Lett 41:4486–4494. https://doi.org/10.1002/2014GL060139

Michael AJ, Field EH, Hardebeck J, Llenos AL, Milner KR, Page MT, Perry SC, van der Elst N, Wein AM (2016) Aftershock forecasting: recent developments and lessons from the 2016 M5.8 Pawnee, Oklahoma earthquake, Abstract S51E-3172 presented at 2016 Fall Meeting, AGU, San Francisco, Calif., 12–16 December

Mileti D, Sorensen J (1990) Communication of emergency public warnings, ORNL-6609. Oak Ridge National Laboratory, Oak Ridge, Tennessee

Mileti DS, Fitzpatrick C, Farhar BC (1992) Fostering public preparations for natural hazards: lessons from the Parkfield earthquake prediction. Environment: Science and Policy for Sustainable Development 34(3):16–39

Mileti DS (1999) Disasters by design: a reassessment of natural hazards in the United States. Joseph Henry Press, Washington, D.C.

Miyazaki SI, McGuire JJ, Segall P (2011) Seismic and aseismic fault slip before and during the 2011 off the Pacific coast of Tohoku Earthquake. Earth, Planets Space 63(7):23

Musson RM, Cecić I (2012) Intensity and intensity scales. New Man Seism Obs Pract 2:1–41

Nakashima T, Okada S, Shinoda A (2017) The importance of seismic death risk assessment of households in the Kumamoto earthquake of 2016. J Disaster Res 12(6):1151–1160

Nanjo KZ, Tsuruoka H, Yokoi S, Ogata Y, Falcone G, Hirata N, Schorlemmer D (2012) Predictability study on the aftershock sequence following the 2011 Tohoku-Oki, Japan, earthquake: first results. Geophys J Int 191(2):653–658

National Institute of Earth Science and Disaster Prevention (NIED) (2012) Seismograph Networks (Hi-net, F-net, Kik-net). http://www.bosai.go.jp/e/research/the_second/earthquake/facilities.html

Ogata Y (1988) Statistical models for earthquake occurrences and residual analysis for point processes. J Am Stat Assoc 83:9–27

Ogata Y (1998) Space-time point-process models for earthquake occurrences. Ann Inst Stat Math 50(2):379–402

Ogata Y, Zhuang J (2006) Space–time ETAS models and an improved extension. Tectonophysics 413(1):13–23

Ohara M, Meguro K, Tanaka A (2011) Survey on people's awareness of earthquake early warning before and after the 2011 Off the Pacific Coast of Tohoku Earthquake. In: Proceedings of the 10th international symposium on new technologies for urban safety of mega cities in Asia, Chiang Mai, Thailand, October 2011, 163–171

Page MT, Van Der Elst N, Hardebeck J, Felzer K Michael AJ (2016) Three ingredients for improved global aftershock forecasts: Tectonic region, time-dependent catalog incompleteness, and intersequence variability. Bulletin Seismol Soc Am 106(5). https://doi.org/10.1785/0120160073

Prado M (2016, September 24) Bay Area earthquake probabilities increase, updated USGS forecast says, The Mercury News, www.mercurynews.com/2016/09/24/bay-area-earthquake-probabilities-increase-updated-usgs-forecast-says/. Accessed 8 Jan 2018

Reasenberg PA (1999) Foreshock occurrence before large earthquakes. J Geophys Res Solid Earth 104(B3):4755–4768

Reasenberg PA, Jones LM (1989) Earthquake hazard after a mainshock in California. Science 243:1173–1176

Reasenberg PA, Jones LM (1994) Earthquake aftershocks: update. Science 265:1251

Rhoades DA, Liukis M, Christophersen A, Gerstenberger MC (2016) Retrospective tests of hybrid operational earthquake forecasting models for Canterbury. Geophys J Int 204(1):440–456

Richter CF (1958) Elementary seismology. W.H. Freeman and Company, San Francisco and London, viii+768 pp

Rikitake T (1979) The large-scale earthquake countermeasures act and the earthquake prediction council in Japan. EOS Trans AGU 60(32):553–555. https://doi.org/10.1029/EO060i032p00553

Roeloffs E, Goltz J (2017) The California earthquake advisory plan: Aa history. Seismol Res Lett 88(3). https://doi.org/10.1785/0220160183

Roeloffs E, Langbein J (1994) The earthquake prediction experiment at Parkfield. California Rev Geophys 32(3):315–336

Segall P, Bradley AM (2012) Slow-slip evolves into megathrust earthquakes in 2D numerical simulations. Geophys Res Lett 39(18). https://doi.org/10.1029/2012gl052811

Stover CW, Coffman JL (1993) Seismicity of the United States, 1568–1989 (Revised). United States geological survey professional paper 1527, U.S. Government Printing Office, Washington D.C

Strader A, Schneider M, Schorlemmer D (2017) Prospective and retrospective evaluation of five-year earthquake forecast models for California. Geophys J Int 211(1):239–251

Uehira K, Kanazawa T, Mochizuki M, Fujimoto H, Noguchi S, Shimbo T, Shiomi K, Kunugi K, Aoi S, Matsumoto T, Sekiguchi S, Okada Y, Shinohara M, Yamada T (2015) Outline of seafloor observation network for Earthquakes and Tsunamis along the Japan Trench (**S**-net). EGU General Assembly 2016(EGU2016-13832):2016

Umeda S (2013) Japan: legal responses to the Great East Japan Earthquake of 2011, The Law Library of Congress, Global Legal Research Center (September), law@loc.gov

United States Geological Survey (2016) ShakeAlert, https://www.shakealert.org/implementation/beta-users/

Wallace RE, Scott S (1999) Connection: the EERI Oral History Series, Robert E. Wallace, https://www.eeri.org/site/images/projects/oralhistory/wallace.pdf. Accessed 4 Jan 2018)

Wang K, Rogers GC (2014) Earthquake preparedness should not fluctuate on a daily or weekly basis. Seismol Res Lett 85:569–571

Wein A, Becker J (2013) Communicating aftershock risk: Roles for reassuring the public, Risk Frontiers 13. https://www.riskfrontiers.com/newsletters/rfnewsVol13_Issue3_Dec13_web.pdf

Wood HO, Neumann F (1931) Modified Mercalli intensity scale of 1931. Bull Seismol Soc Am 21(4):277–283

Working Group on California Earthquake Probabilities (WGCEP) (1988) Probabilities of large earthquakes occurring in California on the San Andreas fault. U.S. Geological Survey, Open-File Report 88-398, 62

Working Group on California Earthquake Probabilities (WGCEP) (1990) Probabilities of large earthquakes in the San Francisco Bay Region, California. U.S. Geological Survey Circular 1053, 51

Working Group on California Earthquake Probabilities (WGCEP) (D. D. Jackson, K. Aki, C. A. Cornell, J. H. Dieterich, T. L. Henyey, M. Mahdyiar, D. Schwartz, S. N. Ward) (1995) Seismic hazards in southern California: probable earthquakes, 1994–2024. Bull Seismol Soc Am 85:379–439

Zechar JD, Schorlemmer D, Liukis M, Yu J, Euchner F, Maechling PJ, Jordan TH (2010) The collaboratory for the study of earthquake predictability perspective on computational earthquake science. Concurr Comput Pract Exp 22(12):1836–1847

MIX
Papier aus verantwortungsvollen Quellen
Paper from responsible sources
FSC® C105338
www.fsc.org

If you have any concerns about our products,
you can contact us on
ProductSafety@springernature.com

In case Publisher is established outside the EU,
the EU authorized representative is:
Springer Nature Customer Service Center GmbH
Europaplatz 3, 69115 Heidelberg, Germany

Printed by Libri Plureos GmbH
in Hamburg, Germany